The CRISPR/Cas System

Emerging Technology and Application

https://doi.org/10.21775/9781910190630

Edited by

Muhammad Jamal

State Key Laboratory of Agricultural Microbiology
College of Animal Science and Medicine
Huazhong Agricultural University
Wuhan
The People's Republic of China

Caister Academic Press

Caister Academic Press
Norfolk, UK

www.caister.com

British Library Cataloguing-in-Publication Data
A catalogue record for this book is available from the British Library

ISBN: 978-1-910190-63-0 (paperback)
ISBN: 978-1-910190-64-7 (ebook)

Cover design adapted from Figure 2.1.

Contents

Current Books of Interest

Illustrated Dictionary of Parasitology in the Post-Genomic Era	2017
Next-generation Sequencing and Bioinformatics for Plant Science	2017
Brewing Microbiology: Current Research, Omics and Microbial Ecology	2017
Metagenomics: Current Advances and Emerging Concepts	2017
Bacillus: Cellular and Molecular Biology (Third Edition)	2017
Cyanobacteria: Omics and Manipulation	2017
Foot-and-Mouth Disease Virus: Current Research and Emerging Trends	2017
Brain-eating Amoebae: Biology and Pathogenesis of *Naegleria fowleri*	2016
Staphylococcus: Genetics and Physiology	2016
Chloroplasts: Current Research and Future Trends	2016
Microbial Biodegradation: From Omics to Function and Application	2016
Influenza: Current Research	2016
MALDI-TOF Mass Spectrometry in Microbiology	2016
Aspergillus and *Penicillium* in the Post-genomic Era	2016
The Bacteriocins: Current Knowledge and Future Prospects	2016
Omics in Plant Disease Resistance	2016
Acidophiles: Life in Extremely Acidic Environments	2016
Climate Change and Microbial Ecology: Current Research and Future Trends	2016
Biofilms in Bioremediation: Current Research and Emerging Technologies	2016
Microalgae: Current Research and Applications	2016
Gas Plasma Sterilization in Microbiology: Theory, Applications, Pitfalls and New Perspectives	2016
Virus Evolution: Current Research and Future Directions	2016
Arboviruses: Molecular Biology, Evolution and Control	2016
Shigella: Molecular and Cellular Biology	2016
Aquatic Biofilms: Ecology, Water Quality and Wastewater Treatment	2016
Alphaviruses: Current Biology	2016
Thermophilic Microorganisms	2015
Flow Cytometry in Microbiology: Technology and Applications	2015
Probiotics and Prebiotics: Current Research and Future Trends	2015
Epigenetics: Current Research and Emerging Trends	2015
Corynebacterium glutamicum: From Systems Biology to Biotechnological Applications	2015
Advanced Vaccine Research Methods for the Decade of Vaccines	2015
Antifungals: From Genomics to Resistance and the Development of Novel Agents	2015
Bacteria-Plant Interactions: Advanced Research and Future Trends	2015

Full details at www.caister.com

Dedication

This book is dedicated to my family, teachers and friends.

Preface

CRISPR/Cas9 (clustered regularly interspaced short palindromic repeats) and CRISPR-associated gene (Cas) form an adaptive immune response of most archaea and bacteria. Initially discovered in 1987 by a Japanese group, the system was characterized in 2002. In 2012, the *in vitro* cleavage activity of Cas9 was observed followed by its first genome engineering application in eukaryotic cells. To date, more than 4000 papers have been published showing the development and applications of Cas9 for wide areas of biological research. The extensive review of the applications of this system across diverse fields is important to explore further avenues and research directions of this highly potent editing technology. For this book, *The CRISPR/Cas System: Emerging Technology and Application*, we invited several researchers to discuss CRISPR/Cas technology across various genomes, from plants to animals, from reproduction to development, from its role in immunity to genetic diseases, from the system structure to system specificity and to debate ethical concerns.

This volume is designed keeping in view the urgent need of scientists to get a rigorous review of the applications and new opportunities in using CRISPR/Cas system to progress the field more efficiently. This volume also deals with some of the potential problems with CRISPR/Cas system that need urgent attention. Specificity of CRISPR/Cas system remains the topic of attention for many researchers across the globe as improvement in this can lead to the more direct and efficient use of CRISPR/Cas system in clinical set-up. The speed and efficiency of the CRISPR/Cas system in *ex vivo* applications make it one of best systems for gene therapy. The coming years will unfold many potential uses of the CRISPR/Cas technology, of which this review volume on its applications will prove a seed of thought.

Muhammad Jamal

Type III CRISPR/Cas System: Introduction and its Application for Genetic Manipulations

1

Tao Liu[1], Saifu Pan[1], Yingjun Li[1,2], Nan Peng[1*] and Qunxin She[2]

[1]State Key Laboratory of Agricultural Microbiology, College of Life Science and Technology, Huazhong Agricultural University, Wuhan, The People's Republic of China.
[2]Archaea Centre, Department of Biology, University of Copenhagen, Copenhagen, Denmark.

*Correspondence: nanp@mail.hzau.edu.cn

https://doi.org/10.21775/9781910190630.01

Abstract

Clustered regularly interspaced short palindromic repeats (CRISPR) and CRISPR-associated (Cas) genes provide adaptive immunity against invasion of foreign nucleic acids in archaea and bacteria. The system functions in three distinct stages: adaptation, biogenesis, and interference. CRISPR/Cas systems are currently classified into at least five different types, each with a signature protein among which Type III systems exhibit a dual DNA/RNA interference activity. Structures of a few Type III surveillance complexes have been determined: they are composed of several different subunits and exhibit striking architectural similarities to Type I surveillance complexes. Here, we review the genetic, biochemical, and structural studies concerning CRISPR/Cas Type III systems and discuss their application for genetic manipulations, including genome engineering and gene silencing.

Introduction

Clustered regularly interspaced short palindromic repeats (CRISPR) and CRISPR-associated (Cas) genes encode the adaptive immune system in prokaryotes, which protects the host against invasion by foreign mobile genetic elements such as virus and plasmid DNA (Barrangou *et al.*, 2007; Brouns *et al.*, 2008; Makarova *et al.*, 2011). CRISPR/Cas systems are widespread and are found in about 90% of archaeal and 40% of bacterial genomes (Grissa *et al.*, 2007; Kunin *et al.*, 2007). In the current classification based on signature protein families and gene synteny of *cas* gene loci, CRISPR/Cas systems comprise two main classes (Classes I and II), which are further divided into at least five different types. Whereas Class I systems (Type II and V) utilize a single Cas protein such as Cas9 to confer immunity, Class I systems employ multiple Cas proteins that form an effector complex to perform the same function (Makarova *et al.*, 2015; Shmakov *et al.*, 2015).

All known CRISPR/Cas systems function in three distinct stages: adaptation, biogenesis, and interference. First, immunity is acquired in a prokaryotic host by integrating short DNA fragments from the genome of an invading virus or plasmid DNA into a CRISPR locus, immediately after the first repeat (Cady *et al.*, 2012; Erdmann *et al.*, 2014; Li *et al.*, 2014; Liu *et al.*, 2015; Richter *et al.*, 2014; Yosef *et al.*, 2012). The CRISPR is transcribed as premature RNA and processed into short mature crRNA by distinct mechanisms, reflected in the diversification of CRISPR/Cas into various subtypes. In Type I and Type III systems, Cas6 family proteins (Carte *et al.*, 2008) or alternatively Cas5d (Garside *et al.*, 2012) catalyses processing within the repeats, and in Type II systems, a trans-acting small RNA directs the pre-crRNA processing by endoribonuclease III and Cas9 within the repeats (Deltcheva *et al.*, 2011). At the interference stage, crRNAs in the effector complexes guide foreign nucleic acids recognition and confer target destruction. A number of recent review articles have been published for other CRISPR/Cas systems and their applications, with emphasis on Type II system (Jamal *et al.*, 2016; Mojica and Montoliu, 2016; Wright *et al.*, 2016). Here we present an overview of the functions of Type III systems and introduce their possible applications in genetic manipulations.

Functions of Type III CRISPR/Cas systems

Classification of Type III CRISPR/Cas systems

The composition and organization of type III CRISPR/Cas loci are diverse, suggesting that these systems could have undergone gene duplications and deletions, domain insertions and fusions during evolution. In addition, a distinct class of Cas-accessory proteins carrying poorly characterized domains have been identified, which could either be involved in crRNA–effector complex functions or in associated immunity (Makarova *et al.*, 2015). Type III CRISPR/Cas systems are characterized by the presence of the signature gene *cas10*, coding for a multidomain protein such as a HD domain and a Palm domain (a variant of the RNA recognition motif) both of which are involved in target nucleic acids interference. Type III CRISPR/Cas systems were first classified into two subtypes: III-A (previously known as Csm module or Mtube subtype) and III-B (also known as Cmr or Cas RAMP module), each of which has a signature protein, i.e. Csm2 for III-A and Cmr5 for III-B subtypes (Makarova *et al.*, 2011). The interference module Type III-A Csm loci usually contain *cas1*, *cas2*, and *cas6* genes, whereas many Type III-B Cmr loci lack these genes (Fig. 1.1), suggesting that they probably share certain functions with other CRISPR/Cas systems in the same genome, such as CRISPR processing and spacer acquisition (Makarova *et al.*, 2013).

Two type III variants (one from subtype III-A and one from subtype III-B) are upgraded to subtypes III-D and III-C, respectively (Makarova *et al.*, 2015; Vestergaard *et al.*, 2014). Both Type III-C and Type III-D lack the *cas1* and *cas2* genes; Type III-C loci encode a Cas10 with inactivated cyclase-like domain, and Type III-D loci typically encode a Cas10 that lacks the HD domain (Makarova *et al.*, 2015) (Fig. 1.1). All type III loci encode small subunit proteins, including one Cas5 protein (Csm4 in Type III-A; Cmr3 in Type III-B and C; and Csx10 in Type III-D) and typically, several Cas7 paralogue proteins (Csm3 and Csm5 in Type III-A; Cmr1, Cmr4, Cmr6 in Type III-B, and III-C; Csm3 in Type III-D) (Makarova *et al.*, 2015; Vestergaard *et al.*, 2014).

crRNA processing

CRISPR arrays are transcribed into long precursors containing spacers and repeats, and are processed into mature small CRISPR RNAs (crRNAs) that act as guides for a targeting complex, which cleaves the genetic elements of the invading virus or plasmid. Biogenesis of mature crRNAs requires a primary cleavage within the repeat sequences to generate intermediates and secondary maturation of these intermediates by additional cleavage. In Type III CRISPR/Cas systems, pre-crRNA processing is initiated by Cas6 endoribonuclease within the repeat sequences, thus generating crRNAs containing complete spacer sequences flanked by an 8-nt repeat sequence at the 5′ end and the rest of the repeat at the other end (Carte *et al.*, 2008; Hatoum-Aslan *et al.*, 2011). In the *Staphylococcus epidermidis* Type III-A system, Cas10 and Csm4 in addition to Cas6 were required for primary processing, to produce an intermediate crRNA of 71 nt. Additional nucleolytic cleavage of the 3′ end generates smaller mature crRNAs and the length of mature crRNA was determined by the 5′ primary processing site (Hatoum-Aslan *et al.*, 2011). Furthermore, Csm2, Csm3, and Csm5 could mediate crRNA maturation following primary processing (Hatoum-Aslan *et al.*, 2011). The crRNA from the *S. solfataricus* Csm complex showed a conserved 8-bp repeat derived sequence at the 5′ end and further processed sequences of 3 bp from the repeat sequences at the 3′ end of the mature crRNA (Rouillon *et al.*, 2013). Similarly, in the *S. solfataricus* P2 Type III-B system, 3′ termini of the crRNA extracted from the Cmr complex showed more variables that displayed

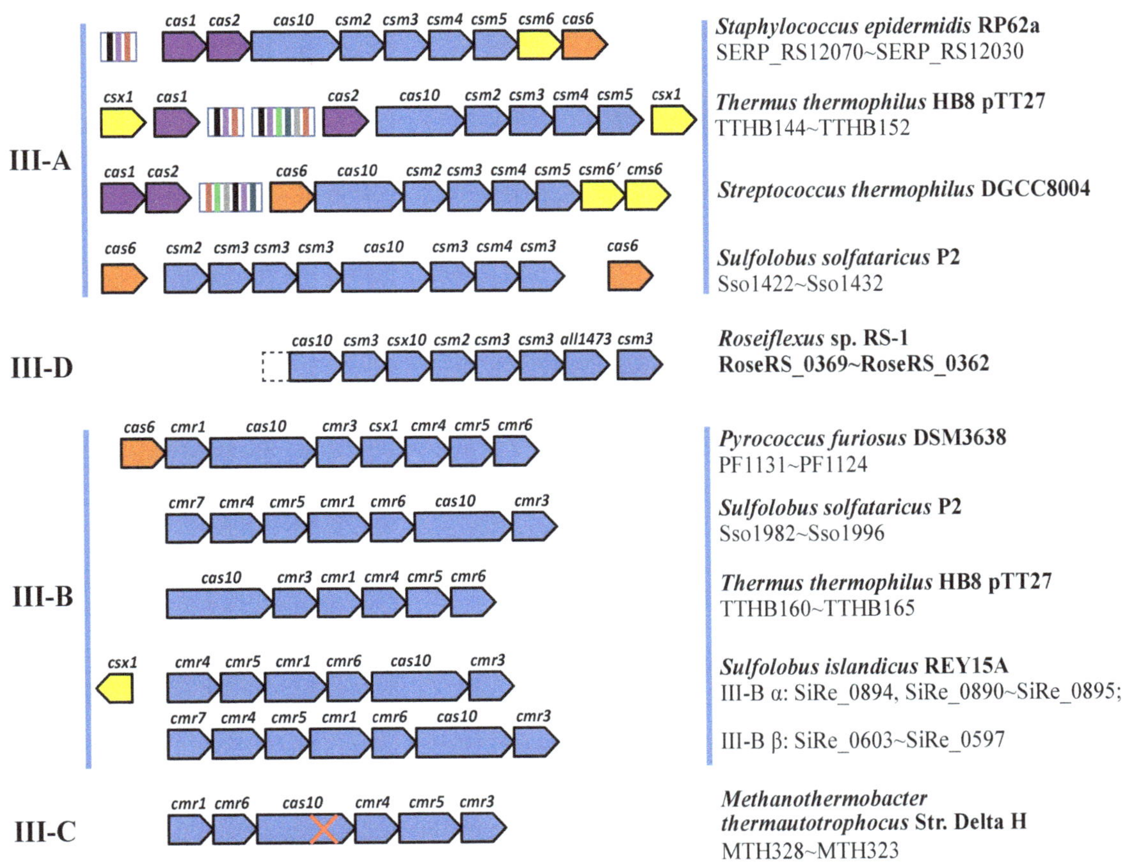

Figure 1.1 Architecture of the genomic loci for the typical Type III CRISPR/Cas systems. Gene organization is shown for the typical subtypes of Type III systems. The representative genomes and the corresponding gene locus tag names are indicated for each subunits, except for the *Streptococcus thermophilus* DGCC8004 strain. Rectangles represent the CRISPR arrays. Genes encoding the adaptation modules are in purple, genes encoding the components of the interference complex are in blue, and the genes encoding the nuclease for CRISPR RNA maturation are in orange. Type III-C loci encode Cas10 with an inactivated cyclase-like domain, and Type III-D loci typically encode Cas10 that lacks the HD domain.

a short handle, or had very few repeat-derived sequences (Zhang *et al.*, 2012). In contrast, crRNA isolated from the *S. solfataricus* P2 Cascade complex still includes the 3′ repeat derived sequence (Lintner *et al.*, 2011), suggesting that crRNAs were differently processed by the Cascade or Cmr complexes. All Type III Csm- or Cmr-bound crRNAs have a variable 3′ end suggesting that processing requires a nuclease for trimming these crRNAs (Hale *et al.*, 2009; Hatoum-Aslan *et al.*, 2011; Rouillon *et al.*, 2013; Staals *et al.*, 2013, 2014; Zhang *et al.*, 2012). Genetic analysis demonstrated that *S. islandicus cmr1a*, *cmr2a*, and *cmr3a* mutant strains lacked the smallest mature crRNA (~38 nt), and that the *cas6* mutant strain lacked both the ~38 nt and ~45 nt mature crRNA, suggesting the presence of secondary processing synergy between Cas6 and Cmr-α interference proteins (Deng *et al.*, 2013).

RNA cleavage activity

CRISPR crRNA-guided RNA cleavage was first reported for a purified native effector complex of the *Pyrococcus furiosus* Type III-B Cmr system (Hale *et al.*, 2009). The *P. furiosus* Cmr complex contains Cmr1, Cmr2 (Cas10), Cmr3, Cmr4, Cmr5, and Cmr6 subunits (Hale *et al.*, 2009). A distinct Type III-B Cmr complex has been purified from *S. solfataricus*, and this effector complex contains seven subunits including Cmr1 to Cmr6, most of which show significant sequence similarity to the corresponding subunits of the *Pyrococcus* Cmr complex and an additional Cmr7 subunit (Zhang *et al.*, 2012). The *Thermus thermophilus* Cmr complex is composed of six different subunits and one crRNA with a stoichiometry of $Cmr1_1 2_1 3_1 4_4 5_3 6_1$:$crRNA_1$ (Staals *et al.*, 2013). The purified Cmr–crRNA complex cleaves complementary RNA but not single-stranded DNA (Hale *et al.*, 2009; Staals *et al.*, 2013; Zhang *et al.*, 2012). The riboendonuclease activity against the RNA complementary to Cmr-bound crRNA requires Mg^{2+} for *T. thermophilus* Cmr complex (Staals *et al.*, 2013) and Mn^{2+} for *S. solfataricus* Cmr complex (Zhang *et al.*, 2012).

The target RNA cleavage by Type III systems shows two distinct patterns. The *S. solfataricus* Cmr complex cleaves target ssRNA complementary to the crRNA at UA sites (Zhang *et al.*, 2012). However, the sizes of the cleavage products were found to be at 6-nt intervals, which has been identified *in vitro* in *T. thermophilus* Type III-A (Staals *et al.*, 2014), *S. thermophilus* Type III-A (Tamulaitis *et al.*, 2014), *P. furiosus* Type III-B (Hale *et al.*, 2014; Osawa *et al.*, 2015; Zhu and Ye, 2015), and *Thermotoga maritima* Type III-B (Estrella *et al.*, 2016), with the substrate initially cleaved at its 3′ end, followed by additional cleavage towards its 5′ end (Staals *et al.*, 2013). Both UA cleavage pattern and 6-nt interval pattern were found *in vivo* in *S. islandicus* Type III-B system (Peng *et al.*, 2015). *T. thermophilus* Cmr complex cleaves ssRNA target initially at its 3′ end, followed by additional cleavage towards to its 5′ end (Staals *et al.*, 2013). In *T. maritime*, target ssRNA cleavage by the Cmr–crRNA complex requires an intact 5′ handle from the repeat sequences of crRNA that cleaves the target at 6-nt intervals (Estrella *et al.*, 2016). In this system, the target was cleaved sequentially, starting predominantly at one site and then proceeding to other sites, and the 3′ end of the crRNA limited the cleavage boundary (Estrella *et al.*, 2016). *S. islandicus* REY15A encodes two type III-B systems (Cmr-α and Cmr-β) and the cleavage patterns clustered into two distinct groups: those with UA–like cleavage resulted from Cmr-β activity, and those of the two defined positions at the 6-nt interval produced by Cmr-α (Peng *et al.*, 2015).

RNA cleavage by the Cmr and Csm complex is not strictly dependent on complete complementarity between the crRNA and the RNA target *in vitro* (Estrella *et al.*, 2016; Staals *et al.*, 2014; Tamulaitis *et al.*, 2014) and *in vivo* (Peng *et al.*, 2015), and the matching or mismatching of 5′ repeat crRNA tag to the target RNA shows no difference in RNA targeting by the *S. epidermidis* Csm complex (Samai *et al.*, 2015). However, spacer mutagenesis identified a trinucleotide seed sequence in the 3′ region of the crRNA that was crucial for RNA targeting (Peng *et al.*, 2015).

The Cas7 family protein Cmr4 or Csm3 act as the catalytic subunit in the Cmr or Csm complex for target RNA cleavage. The structural model (see below) indicates that the Cmr4 helical backbone intercalate between the segments of duplexed crRNA:target RNA, thus distorting the crRNA:target RNA duplex after every 5-bp segment (Osawa *et al.*, 2015; Taylor *et al.*, 2015). The Cmr4 D26A and K46A mutants in *P. furiosus* Cmr complex impaired the target RNA cleavage activity (Benda *et al.*, 2014; Ramia *et al.*, 2014a). The effect of the Cmr4 K46A mutation on target RNA cleavage was accompanied by a loss of crRNA association/Cmr complex formation, indicating that K46 plays a critical role in crRNA interaction (Ramia *et al.*, 2014a). In contrast, the Cmr4 D26A mutation resulted in a specific loss of target RNA cleavages without loss of crRNA association/Cmr complex assembly or target RNA association, indicating that Cmr4 D26 is directly important for target RNA cleavage (Benda *et al.*, 2014; Ramia *et al.*, 2014a). The Csm complex shares similar structure with Cmr complex and the Csm3 subunit was inferred to form the helical backbone (see below). Single residue alanine replacement mutagenesis identified Csm3 is the ribonuclease in Csm complex, and the conserved D33 residue in *S. thermophilus* Csm complex (Kazlauskiene *et al.*, 2016; Samai *et al.*, 2015; Tamulaitis *et al.*, 2014) or D32 residue in *S. epidermidis* Csm complex (Samai *et al.*, 2015) is part of the catalytic site. Although Csm6, a subunit containing a Higher Eukaryotes and Prokaryotes Nucleotide-binding (HEPN) domain, was not found to be part of the Type III-A Cas10–Csm complex (Hatoum-Aslan *et al.*, 2013; Samai *et al.*, 2015; Staals *et al.*, 2014; Tamulaitis *et al.*, 2014), it is required for degradation of phage transcripts with nuclease activity-dead Csm3 that enables Type III-A CRISPR/Cas immunity that targets late expressed genes (Jiang *et al.*, 2016).

DNA cleavage activity

The DNA cleavage activity against cognate target DNA is first identified by genetic analysis. The *S. epidermidis* Type III-A system prevents plasmid conjugation and transformation, indicating that this system possess a DNA targeting activity (Marraffini and Sontheimer, 2008, 2010). The Type III-B DNA cleavage activity was identified by plasmid challenging assay in the archaeon *S. islandicus* REY15A, which encodes two Type III-B interference modules (Deng *et al.*, 2013) (Fig. 1.1). In their study, the challenging plasmid contained a protospacer which lacks the CCN PAM for recognition and cleavage by the Type I-A system and will be transcribed so that the produced transcripts could anneal to crRNA and be susceptible to RNA cleavage. It was assumed that if the protospacer transcript was targeted and cleaved by one or both type III-B interference modules, the transformation efficiency of the plasmid construct would not be impaired significantly. However, the challenging plasmid showed low transformation levels, suggesting the interference of plasmid DNA. Sequence analysis of the surviving transformants revealed that some transformants had mutations located in four different genes exclusive to the *cmr-α* gene cassette (*cmr1a*, *cmr2a*, *cmr3a* and *cmr5a*), and half of the transformants carried an identical 98 kb deletion between the CRISPR loci and the *cmr-α* gene cassette containing two CRISPR-related genes *csx1* and *csm6*; complementation with *csx1* restored interference activity in the deletion mutant (Deng *et al.*, 2013). Thus, *S. islandicus* Cmr-α constitutes the first CRISPR system that exhibits dual targeting of RNA and DNA with a single spacer (Deng *et al.*, 2013; Peng *et al.*, 2015).

The presence of protospacer adjacent motifs (PAMs) on the targets, licenses the DNA targeting in Type I and II systems (Gasiunas *et al.*, 2012; Westra *et al.*, 2013), while the 5′ handle derived from the repeat sequence is the signal for distinguishing self CRISPR DNA to avoid autoimmunity by the Type III system (Deng *et al.*, 2013; Kazlauskiene *et al.*, 2016; Manica *et al.*, 2013; Marraffini and Sontheimer, 2010). Moreover, *Sulfolobus* Type III-B systems require no PAM sequences to confer DNA interference (Deng *et al.*, 2013; Manica *et al.*, 2013). Like their RNA cleavage activity, the DNA cleavage activity of the *Sulfolobus* Cmr complex is not strictly dependent on complete complementarity between the crRNA and the DNA target (Manica *et al.*, 2013). Recently, transcription-dependent DNA targeting by the *P. furiosus* Type III-B Cmr system was found to require a functional rPAM (RNA protospacer-adjacent motif), including NGN, NNG, and NNA, in addition to the crRNA target sequence (Elmore *et al.*, 2016).

Cas10 family proteins are the signature and largest subunits in the Type III complex, and were initially considered to be a nuclease responsible for cleaving target DNA because of their ssDNA-specific cleavage activity (Jung *et al.*, 2015; Ramia *et al.*, 2014b). *P. furiosus* Cas10 alone showed Ni^{2+}-dependent ssDNA nuclease activity *in vitro*, and this activity was significantly attenuated in the Cmr complex *in vitro* (Elmore *et al.*, 2016). Both HD nuclease and Palm domains of the *P. furiosus* Cas10 subunit were required for *in vivo* plasmid DNA cleavage in a transcription-dependent manner (Elmore *et al.*, 2016); however, the GGDD motif in the Palm domain of Cas10 is required for DNA cleavage but not the HD domain in the *S. epidermidis* Type III-A system (Hatoum-Aslan *et al.*, 2014; Ramia *et al.*, 2014b; Samai *et al.*, 2015). Moreover, it was found that the GGDD motif in the Palm domain is involved in binding complementary RNA targets in the *T. maritime* Type III-B Cmr system, that activates ssDNA specific nuclease activity in the HD domain of Cas10 subunit (Estrella *et al.*, 2016). However, only the HD nuclease domain of Cas10 was essential for the DNA cleavage activity of *S. thermophilus* Type III-A Csm complex (Kazlauskiene *et al.*, 2016). The *S. solfataricus* Type III-D, a Cmr7 subunit-containing variant of III-B previously denoted as belonging to Type III-A, cleaves cognate RNA targets with a ruler mechanism *in vitro*, and cleaves plasmid DNA depending on both the cyclase and HD nuclease domains of the Cas10 subunit and independently in the presence of a cognate target sequence *in vitro* (Zhang *et al.*, 2016).

The *in vitro* DNA cleavage activity of Csm or Cmr complex requires target ssRNA. Binding of a target ssRNA activates the *in vitro* ssDNA targeting activity of the *T. maritime* Cmr complex (Estrella *et al.*, 2016) and the *S. thermophilus* Csm complex (Kazlauskiene *et al.*, 2016), and the dsDNA cleavage activity of the *P. furiosus* Cmr complex (Elmore *et al.*, 2016). Cleavage of the ssRNA target and dissociation of the resulting fragments then prevent

DNA cleavage; in addition, with uncleavable RNA targets, the complex gained stronger ssDNA cleavage activity *in vitro* than that of the wild type in *T. maritime* Type III-B system (Estrella *et al.*, 2016). Using an RNA cleavage activity-dead *S. thermophilus* Csm complex, 100% of the ssDNase activity was retained even after 70 min of pre-incubation, suggesting that the Csm complex is trapped in the ssDNase-active state (Kazlauskiene *et al.*, 2016). The ribonucleoprotein of the *S. epidermidis* Type III-A Cas10–Csm complex cleaved DNA at the 3′ flanking side of the target in a transcription-dependent manner in an *E. coli in vitro* transcription system (Samai *et al.*, 2015). Moreover, it was found that the ssDNA nuclease activity of the *T. maritime* Cmr complex was not sequence specific and that the Cmr complex cleaved ssDNA after every thymidine (Estrella *et al.*, 2016). However, the *S. thermophilus* Type III-A Csm complex cleaved the unpaired nucleotides of dsDNA at both strands with almost identical rates, and unpaired single-stranded fragments from 12 to 36 nt were cleaved relatively rapidly, 4–8 nt long were cleaved at a moderate rate, whereas 1–2 nt mismatches were resistant to cleavage (Kazlauskiene *et al.*, 2016).

Structural properties

CRISPR/Cas Type I and Type III systems are mechanistically distinct, in that Type I systems target double-stranded DNA (Hochstrasser *et al.*, 2014; Mulepati and Bailey, 2013; Sinkunas *et al.*, 2013) whereas Type III systems target single-stranded RNA and transcriptional active DNA (Deng *et al.*, 2013; Goldberg *et al.*, 2014; Hale *et al.*, 2009; Peng *et al.*, 2015; Samai *et al.*, 2015; Zhang *et al.*, 2012). However, the structure of the Type III surveillance complex is phylogenetically connected to that of the Type I Cascade complex which has been likened to a seahorse with subunits that represent its head, backbone, belly and tail (Jore *et al.*, 2011; Lintner *et al.*, 2011; Mulepati *et al.*, 2014; Wiedenheft *et al.*, 2011; Zhao *et al.*, 2014). The *T. thermophilus* Type III-B complex contained the Cmr component of $Cmr1_1 2_1 3_1 4_4 5_3 6_1$:$crRNA_1$, and electron microscopic analysis showed a sea worm structure composed of a repeating helical backbone of four Cmr4 subunits that are capped by three Cmr5 subunits juxtaposed with a 20-Å channel or groove for RNA duplex binding, and a 'head' containing Cmr1 and Cmr6 adjoins the other side of the proposed RNA-binding groove (Staals *et al.*, 2013). Near-atomic resolution cryoelectron microscopic reconstructions of the *T. thermophilus* native Type III-B Cmr complex with crRNA further reveal that the thumb-like β-hairpin domains of the Cmr4 subunits and the additional Cmr6 subunit intercalate between the segments of duplexed crRNA:target RNA, thus distorting the crRNA:target RNA duplex after every 5 bp segment and disrupt the formation of an extended A-form double helix, allowing cleavage by the catalytic residues of Cmr4 subunits in the *T. thermophilus* Type III-B Cmr complex (Taylor *et al.*, 2015) (Fig. 1.2).

The *P. furiosus* Type III-B Cmr complex showed similar structure with the *T. thermophilus* Cmr complex where three or four Cmr4 subunits formed the backbone and three Cmr5 subunits formed the belly as revealed by cryoelectron microscopy (Ramia *et al.*, 2014a; Spilman *et al.*, 2013). The Cmr4 and Cmr5 helical core is asymmetrically capped on each end by the Cas10 and Cmr3 at the conserved 5′ crRNA tag sequence and Cmr1 and Cmr6 near the 3′ end of the crRNA (Ramia *et al.*, 2014a; Spilman *et al.*, 2013). Formation of the *P. furiosus* Cmr complex requires both the 5′ tag sequence and 5′-OH group of the crRNA, while deletion or modification of the 5′ tag sequence disrupts binding of the Cas10-Cmr5 subcomplex (Hale *et al.*, 2014). The crystal structure of *P. furiosus* Type III-B Cmr complex shows that similar to the *T. thermophilus* Type III-B Cmr complex, the β-hairpins of three Cmr4 backbone subunits intercalate within the crRNA-target RNA duplex, causing nucleotide displacements with 6-nt intervals and thus periodically place the scissile bonds near the crucial aspartate of Cmr4 (Osawa *et al.*, 2015). In the *P. furiosus* Cmr complex, Cmr3 contacts the 5′ handle sequence of crRNA derived from the repeat sequence and Cmr1 previously found to be located at the head of the complex is absent (Osawa *et al.*, 2015). The crRNA in the Cmr complex flipped out at every sixth nucleotide to form an unwound ribbon-like structure of the duplex. Once the target RNA is captured in a base-complementary manner, the target sites are expelled from the duplex, due to steric clashes with the Cmr4 thumbs. Consequently, the target nucleotides become conformationally labile and are placed into the respective active sites in the Cmr4 (Cas7 family subunit) backbone. The Cmr complex then cleaves it at multiple sites at 6-nt

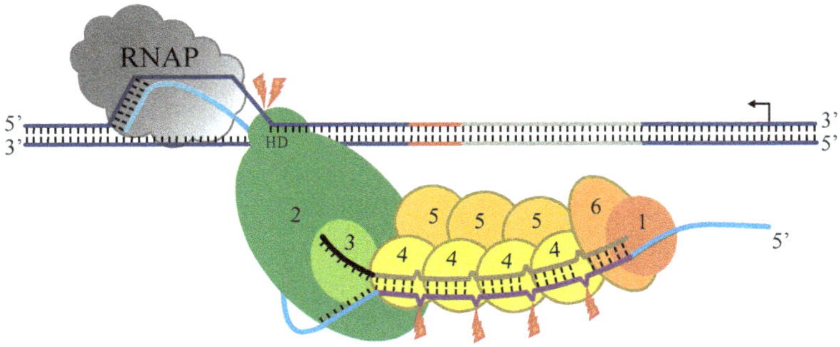

Figure 1.2 Model for RNA targeting and transcription-dependent DNA targeting by Type III CRISPR/Cas system (using the Cmr system as an example). Once RNA Polymerase transcribes through the protospacer, the Type III Cmr complex binds the RNA target mediated by crRNA:target RNA base pairing. The thumb-like β-hairpin domains of the Cmr4 (or Csm3 in Type III-A) subunits and the additional Cmr6 subunit intercalate between the segments of the duplexed crRNA:target RNA, thus distorting the crRNA:target RNA duplex after every 5-bp segment, disrupting the formation of an extended A-form double helix, and allowing cleavage by the catalytic residues of Cmr4 (or Csm3) subunits. The DNA nuclease activity of the HD domain from the Cas10 (Cmr2 or Csm1) subunit was activated by the binding of target RNA and cleaved the non-template DNA strand at the transcription bubble. The mismatch between the 5′ handle of the mature crRNA derived from the CRISPR repeat sequence and the 3′ target RNA sequence is required to avoid the targeting of CRISPR DNA.

intervals, which are specified by the 5′ ruler mechanism that defines the nearest cleavage site to be 5 nt downstream of the 5′ tag (Osawa *et al.*, 2015).

The Csm complex was first purified in *S. solfataricus* and contained Cas10 (Csm1), Csm2, Csm3, and Csm4 subunits (Rouillon *et al.*, 2013), and a Csm5 subunit was identified in the *T. thermophilus* and *S. thermophilus* Type III-A Csm complexes, but Csx1 was not identified (Staals *et al.*, 2014; Tamulaitis *et al.*, 2014). The Csm3 subunit was inferred to form the backbone (Fig. 1.2), Csm2 subunits formed the belly of the smaller filament, and the head of the complexes was capped by Csm5 whereas the foot contained Cas10 (Rouillon *et al.*, 2013; Staals *et al.*, 2014), and the pattern of the cleavage products at 6-nt intervals showed a striking resemblance to those observed with the Type III-B Cmr complex (Staals *et al.*, 2014; Tamulaitis *et al.*, 2014). These data suggest that the Csm complex is related structurally to Type I complexes, sharing a crRNA-binding helical backbone built from the Cas7-family RAMP domain proteins.

Genetic manipulation by Type III CRISPR/Cas system

Genome engineering

CRISPR/Cas system can specifically target DNA sequences by Type I, II, and III systems by crRNA:target nucleotide base-pairing mediated recognition. Among the three main types of CRISPR/Cas systems, DNA targeting by Type II systems requires a single Cas9 nuclease with multiple functions with a trans-acting RNA (tracrRNA) and the mature crRNA (Deltcheva *et al.*, 2011; Gasiunas *et al.*, 2012; Jinek *et al.*, 2012). Because of its simplicity, the Type II CRISPR/Cas9 system has been explored for genome editing immediately after its discovery (Cong *et al.*, 2013; Mali *et al.*, 2013). However, the Type II CRISPR/Cas9 system from mesophilic bacteria may not work in hyperthermophiles, and other CRISPR/Cas subtypes are needed for genetic manipulations in hyperthermophiles, requiring the construction of endogenous CRISPR/Cas systems as genetic tools. Type III CRISPR-Cmr or -Csm systems can target RNA via crRNA:target RNA base pairing, thereby activating DNA targeting activity on the non-template strand (Deng *et al.*, 2013; Estrella *et al.*, 2016; Samai *et al.*, 2015) (Fig. 1.2), thus conferring the possibility for gene silencing and genome editing by this system. Recently, we have employed the Type III-Bα system for genome editing in the hyperthermophilic archaeon, *S. islandicus* REY15A (Li *et al.*, 2016). A mini-CRISPR cassette under the control of an arabinose-inducible promoter allows cloning of any spacer sequences that matching the desired genomic sequences at the BspMI restriction sites in the *Sulfolobus* expression vector (Peng *et al.*, 2012, 2015) (Fig. 1.3). The mini-CRISPR was transcribed and processed by the endogenous Cas6 protein (Peng *et al.*, 2013), and mature crRNA base-pairing, for example, to the stop codon region of the non-coding strand of the *cmr2a* gene; the mismatches between the 5′ tag of crRNA derived from the repeat sequence and the 3′ flanking sequence of targeted sequence on *cmr2a* gene facilitated DNA interference by the Type III-B Cmr-α system in *S. islandicus* Δ*cas3* strain which lacked Type I-A DNA interference activity (Li *et al.*, 2016). Here, the mismatches between the target sequence and the 5′ tag of the mature crRNA derived from the repeat sequence is crucial for licensing the DNA nuclease activity. The DNA interference facilitated homologous recombination between the targeted genomic locus and the donor DNA cloned on the same plasmid resulting in the accurate insertion of a 6×His tag coding sequence at the 3′ end of the *cmr2a* gene. The plasmid carrying mini-CRISPR cassette and the *pyrEF* selection marker can be eliminated by counter-selection after genetic engineering. It seems that the *S. islandicus* Type III-B system did not require a PAM sequence for the discrimination of self and non-self DNA (Deng *et al.*, 2013), even though it is reported that the *P. furiosus* Type III-B Cmr system requires a non-strict rPAM for efficient DNA cleavage (Elmore *et al.*, 2016). Thus, Type III system-mediated genome editing allows the design of spacer (guide) RNAs from any DNA sequence without strict restriction by a PAM. The off-target effect was not determined in the *S. islandicus* Type III-B system; however, off-target effect will kill the cells of *Sulfolobus* hosts

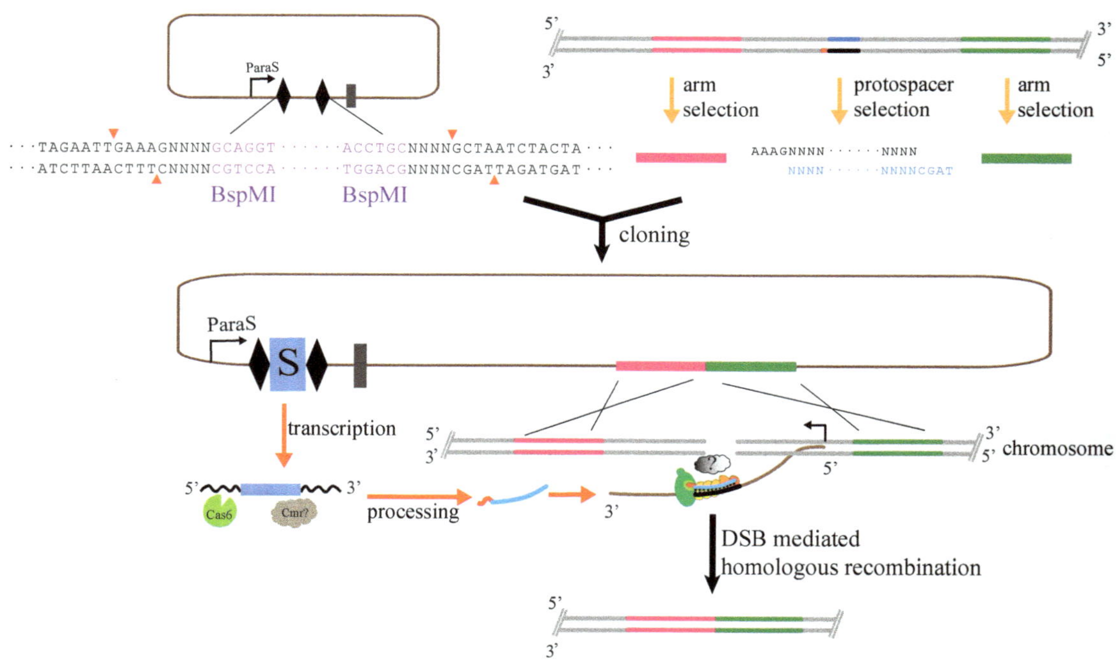

Figure 1.3 Architecture of CRISPR/Cas Type III system mediated genome engineering using the *Sulfolobus* Cmr system as an example. The 40-bp protospacer from the target DNA sequence with its 5′ adjacent sequence mismatches to 3′ segment of the repeat sequence is selected and cloned into the site between two repeats of the mini-CRIPSR through BspMI restriction sites. The mini-CRISPR will be transcribed by a inducible promoter (*araS* promoter) and the transcription will be stopped at a transcription terminator element. The left and right arms of the targeted sequence for homologous recombination are cloned into the same plasmid carrying the mini-CRISPR cassette. The plasmid carrying the mini-CRISPR cassette and the recombination arms is transformed into *Sulfolobus*. The mini-CRISPR is transcribed and processed by endogenous Cas6. The endogenous Cmr or Csm complex cleaves the target RNA mediated by mature crRNA:target RNA base pairing and cleaves the non-template DNA strand activated by the binding of target RNA with the Cmr or Csm complex. A DNA break at the target site enhances homologous recombination between the chromosomal DNA and the arms on the plasmid DNA, resulting in gene deletion, mutation, or insertion.

lacking the non-homologous End Joining pathway. Thus, only the cells repaired by homologous recombination with the donor sequences will survive. Furthermore, Type III Cmr or Csm complexes cannot process the CRISPR RNA itself, and so an endogenous Cas6 protein is required additionally (Carte *et al.*, 2008; Peng *et al.*, 2013).

Gene silencing

Genetic study of gene functions may encounter several problems, one of which is that essential genes cannot be knocked out or mutated. Especially in the hyperthermophilic archaeon that we study, even though we have developed several tools for gene editing in *S. islandicus* REY15A (Deng *et al.*, 2009; Zhang *et al.*, 2010, 2013), the *in vivo* function of essential genes are difficult to study. Type III-B Cmr systems have been shown to mediate PAM-independent RNA cleavage *in vitro* (Hale *et al.*, 2009; Spilman *et al.*, 2013; Staals *et al.*, 2013; Zhang *et al.*, 2012) and *in vivo* (Hale *et al.*, 2012). These results suggest that Type III CRISPR/Cas systems can be used as a gene silencing tool *in vivo*. Recently, gene silencing has been demonstrated in *S. solfataricus* (Zebec *et al.*, 2014) and *S. islandicus* (Peng *et al.*, 2015) using the β-galactosidase gene (*lacS*) as a reporter gene by the endogenous CRISPR Type III-B systems.

Construction of the knock-down plasmid is similar to the construction of the Type III type-based knock-out plasmid (Fig. 1.4). However, since the Type III-B Cmr system shows transcription-dependent DNA nuclease activity along with RNA targeting, the 3′ flanking sequences of the targets (or the ‘protospacer’) must match the 5′ handle tag (the pentanucleotide 5′-GAAAG-3′ or 5′-GAGAC-3′ of the 8-nt repeat handle in *Sulfolobus*) of the mature crRNA to avoid DNA targeting (Deng *et al.*, 2013;

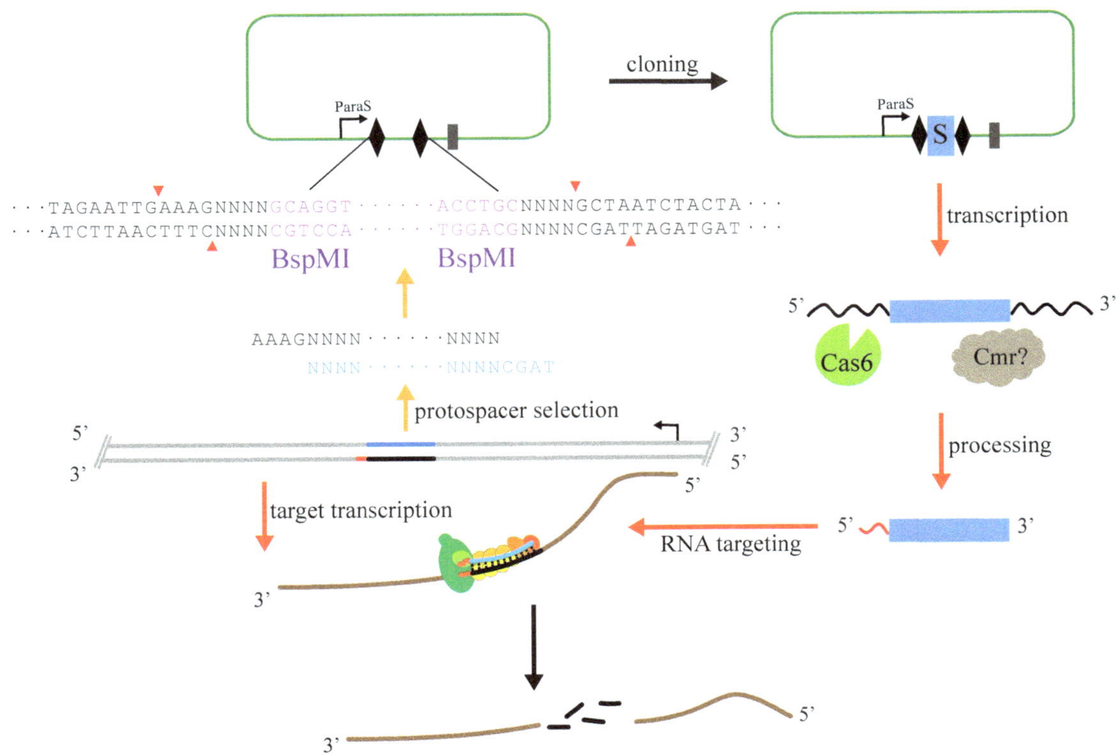

Figure 1.4 Architecture of CRISPR/Cas Type III system mediated gene silencing using the *Sulfolobus* Cmr system as an example. The 40-bp protospacer from the target DNA sequence with its 5′ adjacent sequence matches to 3′ segment of the repeat sequence is selected and cloned into the site between the two repeats of the mini-CRISPR through BspMI restriction sites. The mini-CRISPR will be transcribed by a inducible promoter (*araS* promoter) and processed by endogenous Cas6. The endogenous Cmr or Csm complex cleaves the target RNA mediated by mature crRNA:target RNA base pairing and avoids DNA targeting by base-paring between the 5′ tag of the crRNA and the 3′ adjacent sequence of the RNA target sequence.

Manica *et al.*, 2013). Thus, stretches of 40-nt sequences immediately following the pentanucleotide motifs can be selected as the protospacers to be targeted. However, these motifs will have a limited selection of protospacers from the target sequences for Type III-B-mediated gene silencing. It was reported that two selected spacers matching the middle or 3′ region of the chromosome-borne *lacS* gene coding region reduced ca. 85% of the β-galactosidase activity, but only reduced ca. 40% of the plasmid-borne LacS activity in *S. islandicus* (Peng *et al.*, 2015). This result suggests that Type III-B mediated gene silencing is efficient for the knock-down of expression of single-copy genes or less transcribed genes. In order to reduce the expression of high-copy genes or highly transcribed genes, the transcription of the mini-CRISPR needs to be enhanced by a strong promoter or multiple spacers complementary to the target RNA sequences need to be selected into the mini-CRISPR.

Conclusions

Recently, CRISPR/Cas systems have been classified into two classes (Class I and II) containing five types (Type sI to V). Among these types, the Type III system, further classified into Type III-A and III-B, as well as two less characterized III-C and III-D, is the only system showing both RNA and DNA cleavage activity. All Type III Cmr or Csm complexes have a similar structure where the Cas7 family subunits (Cmr4 in Type III-B systems or Csm3 in III-A systems) form a rigid backbone and the Cmr5 or Csm2 subunits form the belly. The head of the complex is capped by Cmr6/Cmr1 or Csm5, and the foot contains the signature Cas10 (Cmr2 or Csm1) protein and Cmr3. The target RNA is cleaved by Cmr4 or Csm3 mainly at 6-nt intervals, and the non-template ssDNA in the transcription bubble is cleaved by the HD domain and/or cyclase domain of the Cas10 subunit. The sequence-specific RNA targeting and the non-sequence-specific and transcription-dependent DNA targeting facilitates the Type III system-based gene silencing or genome editing tools. However, all well-studied Type III systems are from thermophiles. Characterization of Type III systems from mesophilic bacteria or archaea is important to explore scope of their application in genetic manipulation.

Acknowledgement

We would like to thank the National Natural Science Foundation of China (Nos. 31100050 and 31300042) for financial support.

References

Barrangou, R., Fremaux, C., Deveau, H., Richards, M., Boyaval, P., Moineau, S., Romero, D.A., and Horvath, P. (2007). CRISPR provides acquired resistance against viruses in prokaryotes. Science *315*, 1709–1712. http://dx.doi.org/10.1126/science.1138140

Benda, C., Ebert, J., Scheltema, R.A., Schiller, H.B., Baumgärtner, M., Bonneau, F., Mann, M., and Conti, E. (2014). Structural model of a CRISPR RNA-silencing complex reveals the RNA-target cleavage activity in Cmr4. Mol. Cell *56*, 43–54. http://dx.doi.org/10.1016/j.molcel.2014.09.002

Brouns, S.J., Jore, M.M., Lundgren, M., Westra, E.R., Slijkhuis, R.J., Snijders, A.P., Dickman, M.J., Makarova, K.S., Koonin, E.V., and van der Oost, J. (2008). Small CRISPR RNAs guide antiviral defense in prokaryotes. Science *321*, 960–964. http://dx.doi.org/10.1126/science.1159689

Cady, K.C., Bondy-Denomy, J., Heussler, G.E., Davidson, A.R., and O'Toole, G.A. (2012). The CRISPR/Cas adaptive immune system of *Pseudomonas aeruginosa* mediates resistance to naturally occurring and engineered phages. J. Bacteriol. *194*, 5728–5738. http://dx.doi.org/10.1128/JB.01184-12

Carte, J., Wang, R., Li, H., Terns, R.M., and Terns, M.P. (2008). Cas6 is an endoribonuclease that generates guide RNAs for invader defense in prokaryotes. Genes. Dev. *22*, 3489–3496. http://dx.doi.org/10.1101/gad.1742908

Cong, L., Ran, F.A., Cox, D., Lin, S., Barretto, R., Habib, N., Hsu, P.D., Wu, X., Jiang, W., Marraffini, L.A., *et al.* (2013). Multiplex genome engineering using CRISPR/Cas systems. Science *339*, 819–823. http://dx.doi.org/10.1126/science.1231143

Deltcheva, E., Chylinski, K., Sharma, C.M., Gonzales, K., Chao, Y., Pirzada, Z.A., Eckert, M.R., Vogel, J., and Charpentier, E. (2011). CRISPR RNA maturation by trans-encoded small RNA and host factor RNase III. Nature *471*, 602–607. http://dx.doi.org/10.1038/nature09886

Deng, L., Garrett, R.A., Shah, S.A., Peng, X., and She, Q. (2013). A novel interference mechanism by a type IIIB CRISPR-Cmr module in *Sulfolobus*. Mol. Microbiol. *87*, 1088–1099. http://dx.doi.org/10.1111/mmi.12152

Deng, L., Zhu, H., Chen, Z., Liang, Y.X., and She, Q. (2009). Unmarked gene deletion and host-vector system for the hyperthermophilic crenarchaeon *Sulfolobus islandicus*. Extremophiles *13*, 735–746. http://dx.doi.org/10.1007/s00792-009-0254-2

Elmore, J.R., Sheppard, N.F., Ramia, N., Deighan, T., Li, H., Terns, R.M., and Terns, M.P. (2016). Bipartite recognition of target RNAs activates DNA cleavage by the Type III-B CRISPR-Cas system. Genes. Dev. *30*, 447–459. http://dx.doi.org/10.1101/gad.272153.115

Erdmann, S., Le Moine Bauer, S., and Garrett, R.A. (2014). Inter-viral conflicts that exploit host CRISPR immune systems of *Sulfolobus*. Mol. Microbiol. *91*, 900–917. http://dx.doi.org/10.1111/mmi.12503

Estrella, M.A., Kuo, F.T., and Bailey, S. (2016). RNA-activated DNA cleavage by the Type III-B CRISPR-Cas effector complex. Genes. Dev. *30*, 460–470. http://dx.doi.org/10.1101/gad.273722.115

Garside, E.L., Schellenberg, M.J., Gesner, E.M., Bonanno, J.B., Sauder, J.M., Burley, S.K., Almo, S.C., Mehta, G., and MacMillan, A.M. (2012). Cas5d processes pre-crRNA and is a member of a larger family of CRISPR RNA endonucleases. RNA *18*, 2020–2028. http://dx.doi.org/10.1261/rna.033100.112

Gasiunas, G., Barrangou, R., Horvath, P., and Siksnys, V. (2012). Cas9-crRNA ribonucleoprotein complex mediates specific DNA cleavage for adaptive immunity in bacteria. Proc. Natl. Acad. Sci. U.S.A. *109*, E2579–86. http://dx.doi.org/10.1073/pnas.1208507109

Goldberg, G.W., Jiang, W., Bikard, D., and Marraffini, L.A. (2014). Conditional tolerance of temperate phages via transcription-dependent CRISPR-Cas targeting. Nature *514*, 633–637. http://dx.doi.org/10.1038/nature13637

Grissa, I., Vergnaud, G., and Pourcel, C. (2007). The CRISPRdb database and tools to display CRISPRs and to generate dictionaries of spacers and repeats. BMC Bioinf. *8*, 172. http://dx.doi.org/10.1186/1471-2105-8-172

Hale, C.R., Cocozaki, A., Li, H., Terns, R.M., and Terns, M.P. (2014). Target RNA capture and cleavage by the Cmr type III-B CRISPR-Cas effector complex. Genes. Dev. *28*, 2432–2443. http://dx.doi.org/10.1101/gad.250712.114

Hale, C.R., Majumdar, S., Elmore, J., Pfister, N., Compton, M., Olson, S., Resch, A.M., Glover, C.V., Graveley, B.R., Terns, R.M., *et al.* (2012). Essential features and rational design of CRISPR RNAs that function with the Cas RAMP module complex to cleave RNAs. Mol. Cell *45*, 292–302. http://dx.doi.org/10.1016/j.molcel.2011.10.023

Hale, C.R., Zhao, P., Olson, S., Duff, M.O., Graveley, B.R., Wells, L., Terns, R.M., and Terns, M.P. (2009). RNA-guided RNA cleavage by a CRISPR RNA-Cas protein complex. Cell *139*, 945–956. http://dx.doi.org/10.1016/j.cell.2009.07.040

Hatoum-Aslan, A., Maniv, I., and Marraffini, L.A. (2011). Mature clustered, regularly interspaced, short palindromic repeats RNA (crRNA) length is measured by a ruler mechanism anchored at the precursor processing site. Proc. Natl. Acad. Sci. U.S.A. *108*, 21218–21222. http://dx.doi.org/10.1073/pnas.1112832108

Hatoum-Aslan, A., Maniv, I., Samai, P., and Marraffini, L.A. (2014). Genetic characterization of antiplasmid immunity through a type III-A CRISPR-Cas system. J. Bacteriol. *196*, 310–317. http://dx.doi.org/10.1128/JB.01130-13

Hatoum-Aslan, A., Samai, P., Maniv, I., Jiang, W., and Marraffini, L.A. (2013). A ruler protein in a complex for antiviral defense determines the length of small interfering CRISPR RNAs. J. Biol. Chem. *288*, 27888–27897. http://dx.doi.org/10.1074/jbc.M113.499244

Hochstrasser, M.L., Taylor, D.W., Bhat, P., Guegler, C.K., Sternberg, S.H., Nogales, E., and Doudna, J.A. (2014). CasA mediates Cas3-catalyzed target degradation during CRISPR RNA-guided interference. Proc. Natl. Acad. Sci. U.S.A. *111*, 6618–6623. http://dx.doi.org/10.1073/pnas.1405079111

Jamal, M., Khan, F.A., Da, L., Habib, Z., Dai, J., and Cao, G. (2016). Keeping CRISPR/Cas on-target. Curr. Issues. Mol. Biol. *20*, 1–12.

Jiang, W., Samai, P., and Marraffini, L.A. (2016). Degradation of phage transcripts by CRISPR-Associated RNases enables Type III CRISPR-Cas immunity. Cell *164*, 710–721. http://dx.doi.org/10.1016/j.cell.2015.12.053

Jinek, M., Chylinski, K., Fonfara, I., Hauer, M., Doudna, J.A., and Charpentier, E. (2012). A programmable dual-RNA-guided DNA endonuclease in adaptive bacterial immunity. Science *337*, 816–821. http://dx.doi.org/10.1126/science.1225829

Jore, M.M., Lundgren, M., van Duijn, E., Bultema, J.B., Westra, E.R., Waghmare, S.P., Wiedenheft, B., Pul, U., Wurm, R., Wagner, R., *et al.* (2011). Structural basis for CRISPR RNA-guided DNA recognition by Cascade. Nat. Struct. Mol. Biol. *18*, 529–536. http://dx.doi.org/10.1038/nsmb.2019

Jung, T.Y., An, Y., Park, K.H., Lee, M.H., Oh, B.H., and Woo, E. (2015). Crystal structure of the Csm1 subunit of the Csm complex and its single-stranded DNA-specific nuclease activity. Structure *23*, 782–790. http://dx.doi.org/10.1016/j.str.2015.01.021

Kazlauskiene, M., Tamulaitis, G., Kostiuk, G., Venclovas, Č., and Siksnys, V. (2016). Spatiotemporal Control of Type III-A CRISPR-Cas Immunity: coupling DNA degradation with the target RNA recognition. Mol. Cell *62*, 295–306. http://dx.doi.org/10.1016/j.molcel.2016.03.024

Kunin, V., Sorek, R., and Hugenholtz, P. (2007). Evolutionary conservation of sequence and secondary structures in CRISPR repeats. Genome Biol. *8*, R61. http://dx.doi.org/10.1186/gb-2007-8-4-r61

Li, M., Wang, R., Zhao, D., and Xiang, H. (2014). Adaptation of the *Haloarcula hispanica* CRISPR-Cas system to a purified virus strictly requires a priming process. Nucleic Acids Res. *42*, 2483–2492. http://dx.doi.org/10.1093/nar/gkt1154

Li, Y., Pan, S., Zhang, Y., Ren, M., Feng, M., Peng, N., Chen, L., Liang, Y.X., and She, Q. (2016). Harnessing Type I and Type III CRISPR-Cas systems for genome editing. Nucleic Acids Res. *44*, e34. http://dx.doi.org/10.1093/nar/gkv1044

Lintner, N.G., Kerou, M., Brumfield, S.K., Graham, S., Liu, H., Naismith, J.H., Sdano, M., Peng, N., She, Q., Copie, V., Young, M.J., White, M.F., and Lawrence, C.M. (2011). Structural and functional characterization of an archaeal clustered regularly interspaced short palindromic repeat (CRISPR)-associated complex for antiviral defense (CASCADE). J. Biol. Chem. *286*, 21643–21656. http://dx.doi.org/10.1074/jbc.M111.238485

Liu, T., Li, Y., Wang, X., Ye, Q., Li, H., Liang, Y., She, Q., and Peng, N. (2015). Transcriptional regulator-mediated activation of adaptation genes triggers CRISPR *de novo* spacer acquisition. Nucleic Acids Res. *43*, 1044–1055. http://dx.doi.org/10.1093/nar/gku1383

Makarova, K.S., Haft, D.H., Barrangou, R., Brouns, S.J., Charpentier, E., Horvath, P., Moineau, S., Mojica, F.J., Wolf, Y.I., Yakunin, A.F., *et al.* (2011). Evolution and classification of the CRISPR-Cas systems. Nat. Rev.

Microbiol. *9*, 467–477. http://dx.doi.org/10.1038/nrmicro2577

Makarova, K.S., Wolf, Y.I., Alkhnbashi, O.S., Costa, F., Shah, S.A., Saunders, S.J., Barrangou, R., Brouns, S.J., Charpentier, E., Haft, D.H., *et al.* (2015). An updated evolutionary classification of CRISPR-Cas systems. Nat. Rev. Microbiol. *13*, 722–736. http://dx.doi.org/10.1038/nrmicro3569

Makarova, K.S., Wolf, Y.I., and Koonin, E.V. (2013). The basic building blocks and evolution of CRISPR-CAS systems. Biochem. Soc. Trans. *41*, 1392–1400. http://dx.doi.org/10.1042/BST20130038

Mali, P., Yang, L., Esvelt, K.M., Aach, J., Guell, M., DiCarlo, J.E., Norville, J.E., and Church, G.M. (2013). RNA-guided human genome engineering via Cas9. Science *339*, 823–826. http://dx.doi.org/10.1126/science.1232033

Manica, A., Zebec, Z., Steinkellner, J., and Schleper, C. (2013). Unexpectedly broad target recognition of the CRISPR-mediated virus defence system in the archaeon *Sulfolobus solfataricus*. Nucleic Acids Res. *41*, 10509–10517. http://dx.doi.org/10.1093/nar/gkt767

Marraffini, L.A., and Sontheimer, E.J. (2008). CRISPR interference limits horizontal gene transfer in staphylococci by targeting DNA. Science *322*, 1843–1845. http://dx.doi.org/10.1126/science.1165771

Marraffini, L.A., and Sontheimer, E.J. (2010). Self versus non-self discrimination during CRISPR RNA-directed immunity. Nature *463*, 568–571. http://dx.doi.org/10.1038/nature08703

Mojica, F.J., and Montoliu, L. (2016). On the origin of CRISPR-Cas technology: from prokaryotes to mammals. Trends Microbiol. *24*, 811–820. http://dx.doi.org/10.1016/j.tim.2016.06.005

Mulepati, S., and Bailey, S. (2013). *In vitro* reconstitution of an *Escherichia coli* RNA-guided immune system reveals unidirectional, ATP-dependent degradation of DNA target. J. Biol. Chem. *288*, 22184–22192. http://dx.doi.org/10.1074/jbc.M113.472233

Mulepati, S., Héroux, A., and Bailey, S. (2014). Structural biology. Crystal structure of a CRISPR RNA-guided surveillance complex bound to a ssDNA target. Science *345*, 1479–1484. http://dx.doi.org/10.1126/science.1256996

Osawa, T., Inanaga, H., Sato, C., and Numata, T. (2015). Crystal structure of the CRISPR-Cas RNA silencing Cmr complex bound to a target analog. Mol. Cell *58*, 418–430. http://dx.doi.org/10.1016/j.molcel.2015.03.018

Peng, N., Deng, L., Mei, Y., Jiang, D., Hu, Y., Awayez, M., Liang, Y., and She, Q. (2012). A synthetic arabinose-inducible promoter confers high levels of recombinant protein expression in hyperthermophilic archaeon *Sulfolobus islandicus*. Appl. Environ. Microbiol. *78*, 5630–5637. http://dx.doi.org/10.1128/AEM.00855-12

Peng, W., Feng, M., Feng, X., Liang, Y.X., and She, Q. (2015). An archaeal CRISPR type III-B system exhibiting distinctive RNA targeting features and mediating dual RNA and DNA interference. Nucleic Acids Res. *43*, 406–417. http://dx.doi.org/10.1093/nar/gku1302

Peng, W., Li, H., Hallstrøm, S., Peng, N., Liang, Y.X., and She, Q. (2013). Genetic determinants of PAM-dependent DNA targeting and pre-crRNA processing in *Sulfolobus islandicus*. RNA Biol. *10*, 738–748. http://dx.doi.org/10.4161/rna.23798

Ramia, N.F., Spilman, M., Tang, L., Shao, Y., Elmore, J., Hale, C., Cocozaki, A., Bhattacharya, N., Terns, R.M., Terns, M.P., *et al.* (2014a). Essential structural and functional roles of the Cmr4 subunit in RNA cleavage by the Cmr CRISPR-Cas complex. Cell. Rep. *9*, 1610–1617. http://dx.doi.org/10.1016/j.celrep.2014.11.007

Ramia, N.F., Tang, L., Cocozaki, A.I., and Li, H. (2014b). *Staphylococcus epidermidis* Csm1 is a 3¢-5¢ exonuclease. Nucleic Acids Res. *42*, 1129–1138. http://dx.doi.org/10.1093/nar/gkt914

Richter, C., Dy, R.L., McKenzie, R.E., Watson, B.N., Taylor, C., Chang, J.T., McNeil, M.B., Staals, R.H., and Fineran, P.C. (2014). Priming in the Type I-F CRISPR-Cas system triggers strand-independent spacer acquisition, bi-directionally from the primed protospacer. Nucleic Acids Res. *42*, 8516–8526. http://dx.doi.org/10.1093/nar/gku527

Rouillon, C., Zhou, M., Zhang, J., Politis, A., Beilsten-Edmands, V., Cannone, G., Graham, S., Robinson, C.V., Spagnolo, L., and White, M.F. (2013). Structure of the CRISPR interference complex CSM reveals key similarities with cascade. Mol. Cell *52*, 124–134. http://dx.doi.org/10.1016/j.molcel.2013.08.020

Samai, P., Pyenson, N., Jiang, W., Goldberg, G.W., Hatoum-Aslan, A., and Marraffini, L.A. (2015). Co-transcriptional DNA and RNA cleavage during Type III CRISPR-Cas immunity. Cell *161*, 1164–1174. http://dx.doi.org/10.1016/j.cell.2015.04.027

Shmakov, S., Abudayyeh, O.O., Makarova, K.S., Wolf, Y.I., Gootenberg, J.S., Semenova, E., Minakhin, L., Joung, J., Konermann, S., Severinov, K., *et al.* (2015). Discovery and functional characterization of diverse class 2 CRISPR-Cas systems. Mol. Cell *60*, 385–397. http://dx.doi.org/10.1016/j.molcel.2015.10.008

Sinkunas, T., Gasiunas, G., Waghmare, S.P., Dickman, M.J., Barrangou, R., Horvath, P., and Siksnys, V. (2013). *In vitro* reconstitution of Cascade-mediated CRISPR immunity in *Streptococcus thermophilus*. EMBO J. *32*, 385–394. http://dx.doi.org/10.1038/emboj.2012.352

Spilman, M., Cocozaki, A., Hale, C., Shao, Y., Ramia, N., Terns, R., Terns, M., Li, H., and Stagg, S. (2013). Structure of an RNA silencing complex of the CRISPR-Cas immune system. Mol. Cell *52*, 146–152. http://dx.doi.org/10.1016/j.molcel.2013.09.008

Staals, R.H., Agari, Y., Maki-Yonekura, S., Zhu, Y., Taylor, D.W., van Duijn, E., Barendregt, A., Vlot, M., Koehorst, J.J., Sakamoto, K., *et al.* (2013). Structure and activity of the RNA-targeting Type III-B CRISPR-Cas complex of *Thermus thermophilus*. Mol. Cell *52*, 135–145. http://dx.doi.org/10.1016/j.molcel.2013.09.013

Staals, R.H., Zhu, Y., Taylor, D.W., Kornfeld, J.E., Sharma, K., Barendregt, A., Koehorst, J.J., Vlot, M., Neupane, N., Varossieau, K., *et al.* (2014). RNA targeting by the type III-A CRISPR-Cas Csm complex of *Thermus thermophilus*. Mol. Cell *56*, 518–530. http://dx.doi.org/10.1016/j.molcel.2014.10.005

Tamulaitis, G., Kazlauskiene, M., Manakova, E., Venclovas, Č., Nwokeoji, A.O., Dickman, M.J., Horvath, P., and Siksnys, V. (2014). Programmable RNA shredding by the type III-A CRISPR-Cas system of *Streptococcus thermophilus*. Mol. Cell *56*, 506–517. http://dx.doi.org/10.1016/j.molcel.2014.09.027

Taylor, D.W., Zhu, Y., Staals, R.H., Kornfeld, J.E., Shinkai, A., van der Oost, J., Nogales, E., and Doudna, J.A. (2015). Structural biology. Structures of the CRISPR-Cmr complex reveal mode of RNA target positioning. Science *348*, 581–585. http://dx.doi.org/10.1126/science.aaa4535

Vestergaard, G., Garrett, R.A., and Shah, S.A. (2014). CRISPR adaptive immune systems of Archaea. RNA Biol. *11*, 156–167. http://dx.doi.org/10.4161/rna.27990

Westra, E.R., Semenova, E., Datsenko, K.A., Jackson, R.N., Wiedenheft, B., Severinov, K., and Brouns, S.J. (2013). Type I-E CRISPR-cas systems discriminate target from non-target DNA through base pairing-independent PAM recognition. PLOS Genet. *9*, e1003742. http://dx.doi.org/10.1371/journal.pgen.1003742

Wiedenheft, B., Lander, G.C., Zhou, K., Jore, M.M., Brouns, S.J., van der Oost, J., Doudna, J.A., and Nogales, E. (2011). Structures of the RNA-guided surveillance complex from a bacterial immune system. Nature *477*, 486–489. http://dx.doi.org/10.1038/nature10402

Wright, A.V., Nuñez, J.K., and Doudna, J.A. (2016). Biology and applications of CRISPR systems: harnessing nature's toolbox for genome engineering. Cell *164*, 29–44. http://dx.doi.org/10.1016/j.cell.2015.12.035

Yosef, I., Goren, M.G., and Qimron, U. (2012). Proteins and DNA elements essential for the CRISPR adaptation process in *Escherichia coli*. Nucleic Acids Res. *40*, 5569–5576. http://dx.doi.org/10.1093/nar/gks216

Zebec, Z., Manica, A., Zhang, J., White, M.F., and Schleper, C. (2014). CRISPR-mediated targeted mRNA degradation in the archaeon *Sulfolobus solfataricus*. Nucleic Acids Res. *42*, 5280–5288. http://dx.doi.org/10.1093/nar/gku161

Zhang, C., Guo, L., Deng, L., Wu, Y., Liang, Y., Huang, L., and She, Q. (2010). Revealing the essentiality of multiple archaeal *pcna* genes using a mutant propagation assay based on an improved knockout method. Microbiology *156*, 3386–3397. http://dx.doi.org/10.1099/mic.0.042523-0

Zhang, C., Tian, B., Li, S., Ao, X., Dalgaard, K., Gökce, S., Liang, Y., and She, Q. (2013). Genetic manipulation in *Sulfolobus islandicus* and functional analysis of DNA repair genes. Biochem. Soc. Trans. *41*, 405–410. http://dx.doi.org/10.1042/BST20120285

Zhang, J., Graham, S., Tello, A., Liu, H., and White, M.F. (2016). Multiple nucleic acid cleavage modes in divergent type III CRISPR systems. Nucleic Acids Res. *44*, 1789–1799. http://dx.doi.org/10.1093/nar/gkw020

Zhang, J., Rouillon, C., Kerou, M., Reeks, J., Brugger, K., Graham, S., Reimann, J., Cannone, G., Liu, H., Albers, S.V., *et al.* (2012). Structure and mechanism of the CMR complex for CRISPR-mediated antiviral immunity. Mol. Cell *45*, 303–313. http://dx.doi.org/10.1016/j.molcel.2011.12.013

Zhao, H., Sheng, G., Wang, J., Wang, M., Bunkoczi, G., Gong, W., Wei, Z., and Wang, Y. (2014). Crystal structure of the RNA-guided immune surveillance Cascade complex in *Escherichia coli*. Nature *515*, 147–150. http://dx.doi.org/10.1038/nature13733

Zhu, X., and Ye, K. (2015). Cmr4 is the slicer in the RNA-targeting Cmr CRISPR complex. Nucleic Acids Res. *43*, 1257–1267. http://dx.doi.org/10.1093/nar/gku1355

dCas9: A Versatile Tool For Epigenome Editing

2

Daan J.W. Brocken[1], Mariliis Tark-Dame[2*] and Remus T. Dame[1,3*]

[1]Leiden Institute of Chemistry, Gorlaeus Laboratories, Leiden University, Leiden, The Netherlands.
[2]Swammerdam Institute for Life Sciences, University of Amsterdam, Amsterdam, The Netherlands.
[3]Centre for Microbial Cell Biology, Leiden University, Leiden, The Netherlands.

*Correspondence: m.tark@uva.nl and rtdame@chem.leidenuniv.nl

https://doi.org/10.21775/9781910190630.02

Abstract

The epigenome is a heritable layer of information not encoded in the DNA sequence of the genome, but in chemical modifications of DNA or histones. These chemical modifications, together with transcription factors, operate as spatiotemporal regulators of genome activity. Dissecting epigenome function requires controlled site-specific alteration of epigenetic information. Such control can be obtained using designed DNA-binding platforms associated with effector domains to function as targeted transcription factors or epigenetic modifiers. Here, we review the use of dCas9 as a novel and versatile tool for fundamental studies on epigenetic landscapes, chromatin structure and transcription regulation, and the potential of this approach in basic research in these fields.

Introduction

The epigenome is a layer of information that, together with transcription factors, defines the cell-type-specific gene expression pattern of a genome. By definition, epigenetic information is mitotically and/or meiotically heritable, but not directly encoded in the DNA sequence (Bird, 2007; Berger *et al.*, 2009). Epigenetic information consists of covalent chemical modifications [post-translational modification of histone proteins that include, but are not limited to, methyl, acetyl, phosphoryl and ubiquitin groups (Turner, 2012) and methylation of cytosine bases (Vardimon *et al.*, 1982; Schübeler, 2015)] that alter the structure and physicochemical properties of DNA or DNA-bound histones (Bird, 2002; Kouzarides, 2007). Epigenetic modification states are dynamic by nature and depend on enzymes transferring or removing these modifications (Holliday, 1987; Cubas *et al.*, 1999; Chong and Whitelaw, 2004; Youngson and Whitelaw, 2008). The epigenetic state of a genomic region is determined by combinations of modifications, but how exactly the resulting code is determined remains poorly understood (Gardner *et al.*, 2011). These states correlate with gene expression and chromatin structure, and link the modification pattern of DNA and histones of genomic regions to states of development and differentiation (Holliday, 2006; Henikoff and Shilatifard, 2011; Zhou *et al.*, 2011; Turner, 2012; Smith and Meissner, 2013).

A variety of diseases has been linked to mutations in epigenetic maintenance enzymes or to misregulation of genes following aberrations in the epigenetic code (Kelly *et al.*, 2010; Baylin and Jones, 2011; Plass *et al.*, 2013). Epigenome editing could develop into a tool used to revert aberrations in this code that lead to disease. Due to the lack of knowledge of rare modifications and the combinatorics of DNA and histone modifications, it is unclear by what exact mechanisms epigenetic signals lead to downstream effects on gene regulation and chromatin conformation and whether as yet unknown functional elements exist in genomes. To be able to provide answers to these questions, especially those on causality and the order of concerted processes, epigenetic effectors have been employed to study the implications of modification of the epigenome. Sequence-specific targeting of epigenetic modifiers has been instrumental in understanding the roles of epigenetic modifications on gene regulation. Here, we review current efforts aimed at manipulating the epigenome, focusing on the nuclease-deactivated Cas9 (dCas9) as versatile tool for sequence-specific recruitment of effector proteins.

Epigenome editing

Numerous studies have established correlations between epigenetic state and genome activity (Bernstein *et al.*, 2010; Rivera and Ren, 2013; Roadmaps Epigenomics Consortium *et al.*, 2015). Investigating the epigenome and establishing causal relationships rather than correlations will benefit from site-specific targeting of enzymes involved in establishing, disrupting and maintaining epigenetic states. Ideally, approaches to achieve this aim are readily adaptable to accommodate different target specificity and functionality.

The earliest efforts to edit epigenetic states and determine the effects of alterations therein involved gene knock-out or knock-down of specific epigenetic modifiers (Stancheva and Meehan, 2000; Webster *et al.*, 2005), or the use of compounds inhibiting the activity of these modifiers or their targets (Lyko and Brown, 2005). Such global strategies, due to their pleiotropic impact, do not permit direct specific changes to be distinguished from secondary effects of the perturbation. Epigenetic modifiers – or their isolated functional domains – have been fused to DNA-binding proteins targeted at specific loci. Zinc finger proteins (ZFPs), transcription activator-like effectors (TALEs) and nuclease-deactivated Cas9 (dCas9) have thus been employed successfully as tools for direct control of transcription, and – combined with a domain containing nuclease activity – for genome editing (Gaj *et al.*, 2013). These DNA-binding proteins have also been used as vehicles to target epigenetic modifiers to specific loci. This application has been a major step forward in editing the epigenome (de Groote *et al.*, 2012; Thakore *et al.*, 2016). Below, we describe the three adaptable DNA-binding vehicles currently available for targeted recruitment of effectors at designated genomic loci.

ZFPs

Zinc finger proteins (ZFPs) form a large class of DNA-binding proteins that use coordinated zinc ions to stabilize the typical $\beta\beta\alpha$ fold in their modular DNA-recognition domains. Each zinc finger domain consists of about 30 amino acids, capable of sequence-specific recognition of a 3–4 base pair sequence (Pavletich and Pabo, 1991). The canonical zinc finger protein harbours three such domains. For increased target specificity, synthetic zinc finger proteins generally consist of 4–6 zinc finger domains arranged in tandem. Choo *et al.* (1994) demonstrated in 1994 that it is possible to specifically target an oncogenic gene and reduce its transcription through transcriptional blockage by ZFP. This is the first instance of the use of a designed protein for manipulating gene expression at a defined locus. By targeting the histone methyltransferase (HMT) catalytic core using synthetic ZFPs to an endogenous genomic reporter system the causality of histone modifications in inducing repression of transcription was established (Snowden *et al.*, 2002). Histone 3 Lysine 9 methylation (H3K9me) was found to become enriched throughout a 500–1000 bp region around the target site. This spreading of the H3K9me mark, attributed to HP1, a protein associated with heterochromatin and mediator of gene silencing, resulted in repression of transcription (Snowden *et al.*, 2002). These applications of synthetic ZFPs are key examples of an adaptable DNA-binding platform tool, unique at the time, and laid the basis for the use of ZFP fusions in a wide range of applications (Klug, 2010; Urnov *et al.*, 2010; Gersbach *et al.*, 2014).

TALEs

Transcription activator-like effectors (TALEs) are DNA-binding proteins originating from plant pathogenic bacteria that consist of repeated motifs of 33 or 34 amino acids, with residues 12 and 13 – so-called repeat variable di-residues (RVDs) – in each otherwise conserved repeat recognizing one particular base in double stranded DNA (Boch *et al.*, 2009; Moscou and Bogdanove, 2009; Deng *et al.*, 2012). TALE repeats can – by analogy with ZFPs – be designed in tandem to recognize any DNA sequence of interest, with each repeat recognizing one base via its RVD (Zhang *et al.*, 2011; Mussolino and Cathomen, 2012). Since di-residues specific for each nucleotide have been identified, the rational design of DNA recognition is straightforward (Zhang *et al.*, 2011; Reyon *et al.*, 2012). Nevertheless, target specificity of designed TALEs generally needs to be verified *in vitro* or *in vivo* (Morbitzer *et al.*, 2010; Zhang *et al.*, 2011; Grau *et al.*, 2013; Guilinger *et al.*, 2014). However, also TALEs and TALE–effector fusions verified to be target specific and active in a reporter assay may exhibit different activities when targeted at chromosomal sites, attributed to altered chromatin accessibility (Zhang *et al.*, 2011). Although quite recently discovered, TALEs have proven their value next to ZFPs in gene expression modulation and genome editing (Sanjana *et al.*, 2012; Gaj *et al.*, 2013). TALE fusions have also been used to target epigenetic modifiers to genomic loci. TALEs fused to the synthetic transactivation domain VP64 have been shown to up-regulate transcription of the endogenous pluripotency genes *SOX2* and *KLF4* in human 293FT cells, whereas TALE-VP64 targeted *c-MYC* and *OCT4* genes were unaffected (Zhang *et al.*, 2011). A TALE–TET1 fusion has been shown to decrease methylation of specific CpG dinucleotides in the targeted promoter regions, resulting in increased mRNA expression levels of the targeted genes (Maeder *et al.*, 2013a).

The use of both ZFPs and TALEs has been instrumental in achieving systematic and controlled targeting of effectors to defined genomic loci. However, there are some drawbacks to the use of these DNA-binding proteins. The design of ZFPs to target specific DNA sequences is not as straightforward and modular as suggested above, as side chain–side chain interactions within and between adjacent zinc fingers complicate DNA recognition (Wolfe *et al.*, 2000, 2001), precluding a robust 'recognition code', permitting reliable engineering of highly specific DNA-recognition proteins. Although the DNA recognition via defined RVDs is more modular compared to ZFPs, TALE arrays also suffer from off-target binding (Guilinger *et al.*, 2014; Rogers *et al.*, 2015). In fact, it has been shown that the protein context of a TALE repeat influences the DNA-binding specificity of the array (Rogers *et al.*, 2015). A related drawback is that DNA binding of TALEs (Bultmann *et al.*, 2012) and ZFPs (Choo, 1998; Daniel *et al.*, 2002; Vandevenne *et al.*, 2013) is affected by the presence of methylated cytosines in target sequences, making application in genomic contexts more complicated and less flexible. Independent of these drawbacks limiting application, it is important to realize that both tools require re-design of the protein sequence and validation for individual constructs targeting distinct DNA sequences. This process makes adaptation uncertain as well as time and resource consuming. The capability of employing these proteins in multiplex or high-throughput screening applications is limited by this inflexibility of target adaptation.

dCas9

The CRIPSR-associated protein 9 (Cas9) is an antiviral enzyme of the Type II clustered regularly interspaced short palindromic repeat (CRISPR) adaptive immune system in prokaryotes (Barrangou *et al.*, 2007; Bhaya *et al.*, 2011). The endogenous enzyme introduces double-strand breaks in DNA using two catalytic domains (RuvC and HNH). Two RNA molecules – a CRISPR RNA (crRNA) and a trans-activating crRNA (tracrRNA) – are used to guide Cas9 to its target sequence. A protospacer adjacent motif (PAM) flanking the target sequence recognized by the crRNA acts as an additional determinant in target recognition (Jinek *et al.*, 2012). In biotechnological applications the protein is generally guided by a chimeric single-guide RNA (sgRNA), a hybrid of crRNA and tracrRNA (Fig. 2.1).

Cas9 has been very rapidly and widely adopted (Doudna and Charpentier, 2014) as tool for genome editing (Cho *et al.*, 2013a; Cong *et al.*, 2013; Hwang *et al.*, 2013; Jinek *et al.*, 2013; Mali *et al.*, 2013b). The nuclease-deactivated variant, dCas9 (Fig. 2.1), initially created to establish which catalytic amino acids are necessary for dsDNA cleavage (Jinek *et al.*, 2012), laid the foundation for a whole range of new applications needing site-specific targeting. The dCas9 protein has been adopted as DNA-binding platform for applications as diverse as transcriptional blockage (Bikard *et al.*, 2013; Qi *et al.*, 2013), gene expression modulation (Cheng *et al.*, 2013; Gilbert *et al.*, 2013), epigenetic editing (Hilton *et al.*, 2015; Kearns *et al.*, 2015) and staining of chromosomal regions for live cell imaging (Chen *et al.*, 2013; Anton *et al.*, 2014; Ma *et al.*, 2015). In the next sections we discuss in detail

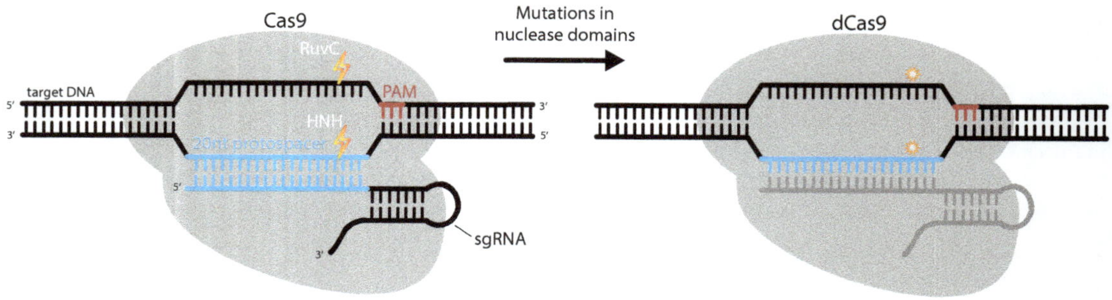

Figure 2.1 The Cas9 endonuclease and its nuclease-deactivated variant dCas9. The CRISPR-associated protein 9 (Cas9) is a prokaryotic antiviral protein that is guided by two RNA molecules (not shown) or, alternatively, one chimeric single-guide RNA (sgRNA) to cleave (using its RuvC and HNH catalytic domains) a target sequence containing a proper protospacer adjacent motif (PAM, in red) and matching the 20 nucleotide protospacer (in blue). Mutations in the nuclease domains (D10A and H840A in the commonly used *Streptococcus pyogenes* Cas9) result in a Cas9 variant, referred to as dCas9, capable of binding to the target sequence, but unable to cleave its target.

design strategies and applications of dCas9 for gene expression modulation and epigenetic editing.

dCas9 design strategies

Different strategies of varying complexity and efficacy for the use of dCas9 to manipulate the epigenome in controlled manner have been developed. In essence all these strategies belong to one of three categories discussed below (Fig. 2.2). Notably, some strategies discussed below have only been used for Cas9, but are equally suitable for use with dCas9.

Class I – direct effector fusion

In the most straightforward design, the dCas9 protein (with no effector fused) is used to interfere with transcription via steric blockage of RNA polymerase binding or transcription elongation (Bikard *et al.*, 2013; Qi *et al.*, 2013). This strategy is successful in prokaryotes leading up to ~300-fold mRNA reduction when a single sgRNA is used to target dCas9 or even ~1000-fold when two sgRNAs are combined to block transcription elongation (Qi *et al.*, 2013). However, interference with transcription has not been successful in mammalian cells yielding only ~2-fold reduction in transcript levels (Qi *et al.*, 2013). For potent modulation of gene expression in mammalian cells by dCas9, specific effectors are implied. Down-regulation (CRISPRi) or activation (CRISPRa) of targeted genes is achieved by genetically fusing effector proteins – or their active domains – to dCas9 and expressing them as a single recombinant protein (Fig. 2.2A). Transcription activator domains (VP64, p65) or repressor domains (KRAB, SID) have been fused to dCas9 to specifically increase or decrease target gene expression (Gilbert *et al.*, 2013; Maeder *et al.*, 2013b; Perez-Pinera *et al.*, 2013; Lawhorn *et al.*, 2014). Single dCas9–effector fusions are commonly targeted to adjacent sites using multiple different sgRNAs for maximum impact. In an effort to obtain maximum activation a combination of three different effectors (VP64, p65 and Rta) has been fused in succession to dCas9, resulting in a ~100-fold increase in transactivation of the target genes compared to dCas9-VP64 alone (Chavez *et al.*, 2015). A basic strategy towards achieving temporal control of dCas9–effector target binding is inducible expression of dCas9 or the sgRNA, e.g. by using a doxycycline (González *et al.*, 2014; Wang *et al.*, 2014; Dow *et al.*, 2015) or IPTG-responsive promoter. However, this approach suffers from leaky expression in the absence of an inducer.

Class II – indirect effector recruitment

Class II strategies incorporate into the basic design additional motifs that recruit effector proteins (Fig. 2.2B). An example of such a motif is the SunTag (Tanenbaum *et al.*, 2014), a protein scaffold containing peptide epitopes (a variant with 10 and one with 24 epitopes is available) able to recruit effector domains via specific single-chain variable fragment (scFv) antibodies (Fig. 2.2B). The SunTag-carrying

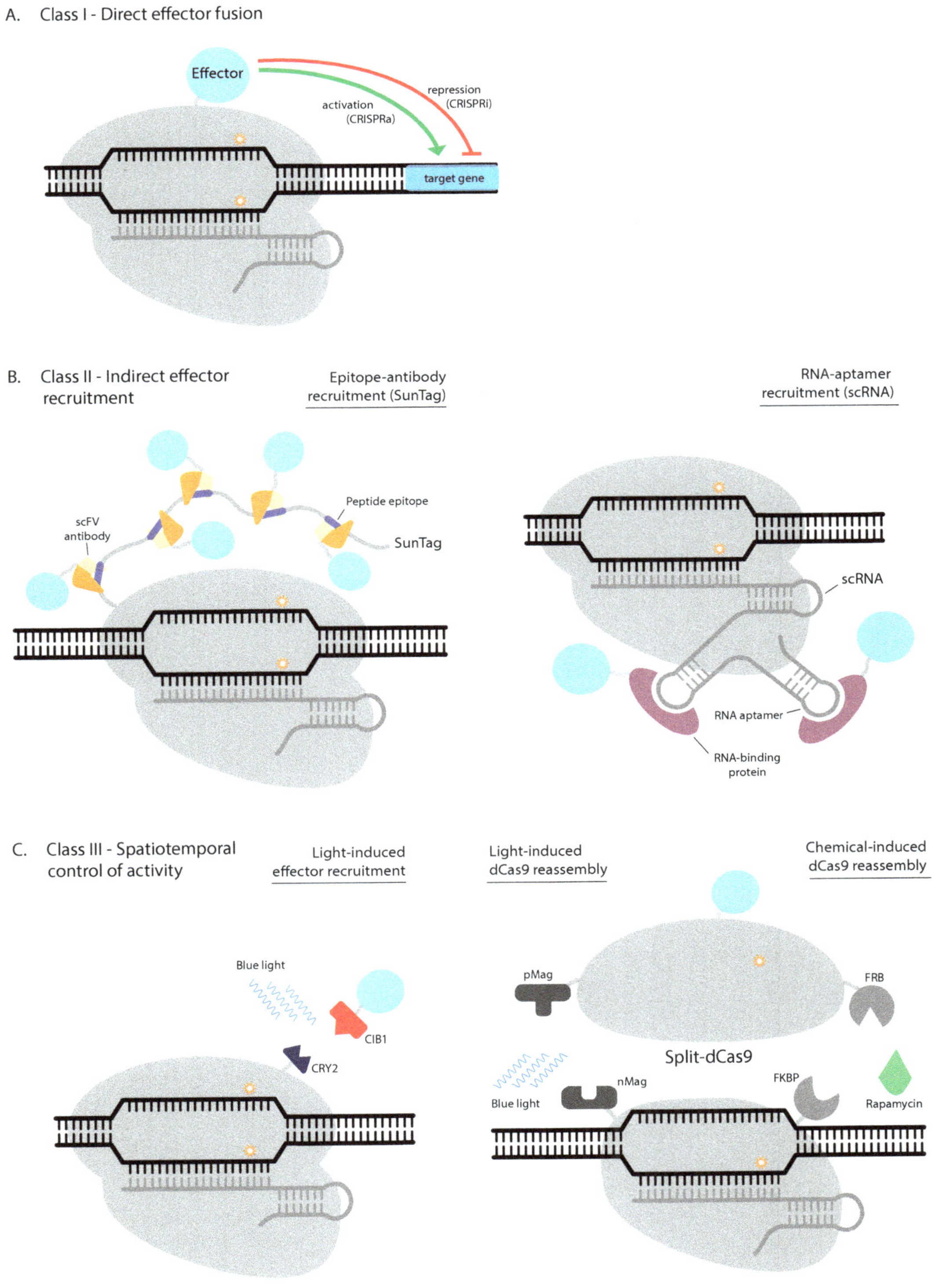

Figure 2.2 Design strategies of using dCas9 to target effector domains to specific DNA sequences. (A) The effector domain is directly fused to dCas9 to recruit it to sequences specified by the sgRNA. (B) The effector domain(s) is recruited via functional scaffolds incorporated in the sgRNA–dCas9 complex, either via fusion to dCas9 (left) or via RNA aptamers in a scaffolding RNA (scRNA, right). (C) Spatiotemporal control of effector activity is obtained via controlled recruitment of effectors to the sgRNA–dCas9 complex (left) or the reconstitution of split-dCas9 directly fused to effectors (right) via light- or chemical-inducible heterodimerization partners.

dCas9 has been shown to successfully recruit scFv–VP64 fusions and to increase target gene expression more than the simple dCas9–VP64 fusion (Tanenbaum *et al.*, 2014).

More complex systems with added functionality are not limited to direct genetic fusion of effector domains to dCas9. The sgRNA can be extended to include RNA aptamers (Fig. 2.2B), secondary RNA structures specifically recognized by RNA-binding proteins, to form a scaffolding RNA (scRNA) (Mali *et al.*, 2013a; Konermann *et al.*, 2015; Zalatan *et al.*, 2015). Using scRNAs with RNA aptamers such as MS2, PP7, com or the PUF binding site (PBS), effectors can be recruited to the dCas9–sgRNA complex indirectly via fusion to corresponding RNA-binding proteins (Mali *et al.*, 2013a; Zalatan *et al.*, 2015; Cheng *et al.*, 2016). Recruiting effectors simultaneously via a dCas9 gene fusion and via aptamers present in the sgRNA has been shown to yield strong synergistic transactivation (Konermann *et al.*, 2015; Xu *et al.*, 2016). For example, recruiting two MS2–p65-HSF1 fusions to dCas9-VP64 via an scRNA containing two MS2 hairpin aptamers resulted in a 100-fold enhancement of transactivation compared to just dCas9-VP64. Using just one scRNA this system outperformed a pool of eight sgRNAs targeting dCas9-VP64 to distinct sites along the proximal promoter region of target genes (Konermann *et al.*, 2015).

Class III – spatiotemporal control of activity

In a third class, dCas9 strategies are aimed at precisely controlling effector recruitment or dCas9 DNA-binding activity in space and/or time (Fig. 2.2C). Such control has been achieved by using split-(d)Cas9 or split-(d)Cas9-effector proteins that are conditionally assembled into a functional DNA-binding complex in the presence of sgRNA (Wright *et al.*, 2015), upon chemical induction (Zetsche *et al.*, 2015b) or light induction (Nihongaki *et al.*, 2015a). (d)Cas9, split in two individual domains, is assembled into a functional complex in the presence of full-length sgRNA. Truncated sgRNA prevents complex formation and can be used to disassemble the two Cas9 domains (Wright *et al.*, 2015). For chemical induction C- and N-terminal Cas9 (and dCas9) fragments have been fused to FK506 binding protein 12 (FKBP) and FKBP rapamycin binding (FRB) domains, respectively, that dimerize upon rapamycin induction (Banaszynski *et al.*, 2005; Zetsche *et al.*, 2015b) (Fig. 2.2C). Using a split-dCas9 system fused to VP64, expression of target genes was shown to be specifically induced in the presence of rapamycin (Zetsche *et al.*, 2015b). Another strategy involves reassembly of split-dCas9 by photoinducible dimerization domains termed Magnets (pMag and nMag) (Fig. 2.2C) in response to exposure to blue light (Kawano *et al.*, 2015; Nihongaki *et al.*, 2015a).

In order to control effector recruitment rather than split-(d)Cas9 reconstitution, the heterodimerization partners CRY2 and CIB1 have been used to bring together full-length dCas9 and effectors such as VP64 or p65 upon blue light irradiation (Konermann *et al.*, 2013; Nihongaki *et al.*, 2015b; Polstein and Gersbach, 2015) (Fig. 2.2C). Although the light-inducible CRY2–CIB1 pair works well to bring together dCas9 and effectors, the use of these partners was unsuccessful when applied to reassembly of split-(d)Cas9 (Nihongaki *et al.*, 2015a).

An alternative strategy to control the binding of Cas9 to its target sequence relies on intein-mediated splicing (Davis *et al.*, 2015; Truong *et al.*, 2015). Designs with intein in the reading frame of full-length Cas9 or fused to both fragments of split-Cas9 have been used. In both cases, after intein trans-splicing is induced full-length functional Cas9 is obtained (Davis *et al.*, 2015; Truong *et al.*, 2015). An other, less common approach, is to use Cas9 inactive due to caging of lysine residues necessary for Cas9 function (Hemphill *et al.*, 2015). Exposure to UV light removes the caging group (Riggsbee and Deiters, 2010) and recovers an active Cas9 (Hemphill *et al.*, 2015).

Due to the availability of Cas9 orthologues (Chylinski *et al.*, 2014) and redesigned synthetic Cas9 proteins (Kleinstiver *et al.*, 2015; Hirano *et al.*, 2016) with different PAM recognition sequences (Table 2.1), combinations of multiple dCas9 orthologues and matching sgRNAs can be used in parallel to perform distinct activities. In *E. coli* two orthologues have been used to simultaneously cleave viral dsDNA and repress a reporter gene via transcriptional blockage (using a nuclease-active SpyCas9 and a deactivated NmCas9 protein respectively) (Esvelt *et al.*, 2013). In this study it was established that *N. meningitidis*, *S. thermophilus* CRISPR1 and *S. pyogenes* Cas9 function fully orthogonal to one another in *E. coli* as well as in

Table 2.1 Overview of Cas9 orthologues and synthetic redesigned Cas9s. Orthologues indicated with a $^{+}$ have been used as dCas9 systems. Abbreviations: N = any base; D = A, G or T; Y = C or T; R = G or A; W = A or T. Note: PAM sequences shown here are consensus sequences derived from sequence logos (i.e. nucleotide frequency plots) retrieved from bioinformatics analyses and/or cleavage assays using libraries of putative PAMs

Cas9	Origin	5′-PAM-3′	References
BlCas9	*Brevibacillus laterosporus*	NNNNCNDD	Karvelis *et al*., 2015
CdCas9	*Corynebacterium diphtheriae*	NGG	Ran *et al*., 2015
CjCas9	*Campylobacter jejuni*	NNNNACA	Fonfara *et al*., 2014
ClCas9	*Campylobacter lari*	NNGGG	Ran *et al*., 2015
FnCas9	*Francisella novicida*	NG	Fonfara *et al*., 2014
FnCas9$_{RHA}$	Mutant E1369R/E1449H/R1556A	YG	Hirano *et al*., 2016
LiCas9	*Listeria innocua*	NGG	Esvelt *et al*., 2013
NcCas9	*Neisseria cinerea*	NNNNGTA	Ran *et al*., 2015
NmCas9^{+}	*Neisseria meningitidis*	NNNNGATT	Hou *et al*., 2013; Zhang *et al*., 2013
PlCas9	*Parvibaculum lavamentivorans*	NNNCAT	Ran *et al*., 2015
PmCas9	*Pasteurella multocida*	GNNNCNNA	Fonfara *et al*., 2014
SaCas9	*Staphylococcus aureus*	NNGRR(T)	Kleinstiver *et al*., 2015; Ran *et al*., 2015
SmCas9	*Streptococcus mutans*	NGG	Fonfara *et al*., 2014; van der Ploeg, 2009
SpaCas9	*Streptococcus pasteurianus*	NNGTGA	Ran *et al*., 2015
SpyCas9^{+}	*Streptococcus pyogenes*	NGG	Jinek *et al*., 2012
SpyCas9$_{EQR}$	Mutant D1135E/R1335Q/T1337R	NGG (increased specificity)	Kleinstiver *et al*., 2015
SpyCas9$_{QQR1}$	Mutant G1218R/N1286Q/I1331F/D1332K/R1333Q/R1335Q/T1337R	NAAG	Anders *et al*., 2016
SpyCas9$_{VQR}$	Mutant D1135V/R1335Q/T1337R	NGAN	Kleinstiver *et al*., 2015
SpyCas9$_{VRER}$	Mutant D1135V/G1218R/R1335E/T1337R	NGCG	Kleinstiver *et al*., 2015
St1Cas9^{+}	*Streptococcus thermophilus*** (CRISPR 1)	NNARAAW	Deveau *et al*., 2008; Fonfara *et al*., 2014; Horvath *et al*., 2008
St3Cas9	*Streptococcus thermophilus** (CRISPR 3)	NGG(NG)	Fonfara *et al*., 2014; Horvath *et al*., 2008
TdCas9^{+}	*Treponema denticola*	NAAAAN	Esvelt *et al*., 2013

human cells, allowing for targeting distinct and non-overlapping sets of sequences within the same cell (Esvelt *et al.*, 2013). The three catalytically deactivated Cas9 orthologues from the same three species have been used to label distinct chromosomal loci in live human cells (Ma *et al.*, 2015).

Applications of the dCas9 tool

Targetable DNA-binding proteins with additional activity in epigenetic editing and/or gene expression modulation can be divided in two categories based on the nature of the effector: (i) targeted transcription factors (TTFs) and (ii) targeted epigenetic modifiers (TEMs) (Table 2.2). Effectors used in TTFs are (derivatives of) natural abundant and potent transcription factors that affect gene expression, but also often have indirect and multilateral effects on the states of the epigenome through recruitment of (multiple) effector partners. Commonly used effectors are herpes simplex viral protein 16, VP16 (Wang *et al.*, 2000; Memedula and Belmont, 2003) – better known in tetrameric form as VP64 (Beerli *et al.*, 1998) – and p65, a subunit of the human NF-κB transcription factor (Schmitz and Baeuerle, 1991). Both effectors exhibit complex mechanisms of transactivation via recruitment of secondary transcription factors (Mittler *et al.*, 2003;

Table 2.2 An overview of effectors used in dCas9 systems

	Direct fusion	Indirect recruitment: scRNA	SunTag	Light-induced	References
Targeted transcription factors (TTFs)					
Activating	VP64				Chavez *et al.*, 2015; Farzadfard *et al.*, 2013; Gao *et al.*, 2014; Gilbert *et al.*, 2013; Hilton *et al.*, 2015; Hu *et al.*, 2014; Kearns *et al.*, 2014; Maeder *et al.*, 2013b; Mali *et al.*, 2013b; Perez-Pinera *et al.*, 2013; Polstein and Gersbach, 2015; Zetsche *et al.*, 2015a
		VP64			Cheng *et al.*, 2016; Mali *et al.*, 2013b; Zalatan *et al.*, 2015
			VP64		Gilbert *et al.*, 2014; Tanenbaum *et al.*, 2014
				VP64	Konermann *et al.*, 2013; Nihongaki *et al.*, 2015b; Polstein and Gersbach, 2015
	VP48				Cheng *et al.*, 2013
	VP120				Cheng *et al.*, 2013; Gao *et al.*, 2014
	VP64	VP64 or p65 +/- HSF1 or MyoD1			Konermann *et al.*, 2015
	p65				Gilbert *et al.*, 2013
	p65	p65 or VP64			Konermann *et al.*, 2015
				p65	Nihongaki *et al.*, 2015b
	P65-HSF1				Cheng *et al.*, 2016
		P65-HSF1			Cheng *et al.*, 2016
	VP64-p65-Rta				Chavez *et al.*, 2015
Repressing	KRAB				Farzadfard *et al.*, 2013; Gao *et al.*, 2014; Gilbert *et al.*, 2013; Hu *et al.*, 2014; Kearns *et al.*, 2014, 2015; Lawhorn *et al.*, 2014; Thakore *et al.*, 2015
		KRAB			Cheng *et al.*, 2016; Zalatan *et al.*, 2015
	SID4x				Konermann *et al.*, 2013
	Mxi1				Gilbert *et al.*, 2013
Targeted epigenetic modifiers (TEMs)					
Histone acetylation	p300				Hilton *et al.*, 2015
	CBP				Cheng *et al.*, 2016
		CBP			Cheng *et al.*, 2016
Histone demethylation	LSD1				Kearns *et al.*, 2015
DNA methylation	DNMT3A				McDonald *et al.*, 2016; Vojta *et al.*, 2016
DNA demethylation	TET1	TET1			Xu *et al.*, 2016

van Essen *et al.*, 2009). As transcription repressor the Krüppel-associated box (KRAB), a domain found in numerous mammalian repressors, is commonly used. Together with recruited co-repressor KAP1, KRAB attracts a variety of epigenetic modifiers and chromatin remodelling proteins to induce heterochromatin and block transcription (Groner *et al.*, 2010). TEMs are designed based on epigenetic effectors, i.e. (derivatives of or active domains of) enzymes that catalyse the transfer or removal of epigenetic modifications. Epigenetic effectors that have been used as TEMs are DNA methyltransferase 3A (DMT3A) (McDonald *et al.*, 2016; Vojta *et al.*, 2016), TET1 5mC-hydroxylase (Maeder *et al.*, 2013a; Xu *et al.*, 2016), LSD1 histone demethylase (Mendenhall *et al.*, 2013; Kearns *et al.*, 2015) and the p300 and CBP histone acetyltransferases (Hilton *et al.*, 2015; Cheng *et al.*, 2016). Below, we discuss applications of epigenetic editing by TEMs or indirect modulation by TTFs using dCas9 as DNA-binding platform.

Manipulating the epigenome and gene expression using TEMs and TTFs

So far only a few studies have been published reporting the use of dCas9 fusions for epigenome editing and manipulating gene expression. The catalytic histone acetyltransferase (HAT) core domain p300 has been used to catalyse acetylation of histones in human HEK293T cells (Hilton *et al.*, 2015). Targeting dCas9–p300 fusions to promoter regions or proximal or distant enhancers caused activation of gene expression. Increased expression upon enhancer-targeting was concomitant with enrichment in H3K27ac at the corresponding genomic target sites (Hilton *et al.*, 2015). In most cases the same genes could be transactivated by dCas9-VP64 when targeted at promoters. To achieve transactivation both effectors can thus be used. The two effectors behave somewhat differently in terms of their impact on histone acetylation state, as p300 directly catalyses H3K27ac (Ogryzko *et al.*, 1996; Delvecchio *et al.*, 2013), whereas VP64 recruits subsequent transactivation components, amongst which is p300 (Memedula and Belmont, 2003). Also the histone acetyltransferase domain of the CREB-binding protein has been fused to dCas9 (dCas9-CBPHAT) and has been used to catalyse locus-specific acetylation of histones (Cheng *et al.*, 2016). dCas9-CBPHAT was targeted using the Casilio (CRISPR/Cas9-Pumilio) system, which harbours an scRNA containing multiple PUF binding sites (PBS), to recruit additional CBPHAT domains via fusions with Pumilio/FBF (PUF) RNA-binding domains. Similar to dCas9-p300, targeting dCas9-CBPHAT to promoters or proximal and distal enhancer caused increased expression of the target genes (Cheng *et al.*, 2016).

dCas9 has also been used to introduce DNA methylation by targeting the catalytic domain of the *de novo* DNA methyltransferase 3A (DNMT3A) to specific loci. dCas9-DNMT3A has been used to methylate cytosines around the transcription start site (TSS) of human and mice genes (McDonald *et al.*, 2016; Vojta *et al.*, 2016). Using this strategy, the methylation status of the CpG island spanning the TSS of the tumour-suppressor gene *CDKN2A* increased by 20%, resulting in 40% reduction in *CDKN2A* mRNA levels (McDonald *et al.*, 2016). Targeting the unmethylated promoter regions of the *BACH* or *IL6ST* genes increased CpG methylation by 50% and reduced expression 2-fold (Vojta *et al.*, 2016). However, this increase in methylation is not the only cause of reduced expression; catalytically inactive DNMT3A also induced repression, attributed to steric hindrance of the transcription machinery (McDonald *et al.*, 2016; Vojta *et al.*, 2016). In contrast, when dCas9-DNMT3A was targeted to a CpG island 100–400 bp downstream of the TSS, resulting in an increase of 20–35% methylation of CpG residues, no significant change in *CDKN2A* mRNA expression was observed (McDonald *et al.*, 2016). Upon targeting dCas9-DNMT3A to the unmethylated *Cdkn1a* gene in mouse myeloid progenitor cells resulting in *Cdkn1a* repression, proliferation was enhanced. In these studies dCas9 was as effective in introducing DNA methylation as ZFP- or TALE-based DNMT3A systems. The highest efficiency of methylation of targeted CpG residues was obtained when target sites are bracketed by inwardly directed sgRNAs (i.e. target CpGs are downstream of the PAM) and within ~50 bp of the sgRNA binding site (McDonald *et al.*, 2016; Vojta *et al.*, 2016).

Recently the catalytic domain of TET1 was used to induce demethylation of DNA (Xu *et al.*, 2016). In this study the catalytic domain of TET1 was fused to dCas9 and two additional copies of the same effector were directed to the target location via two MS2 aptamers integrated in the sgRNA design (Xu *et al.*, 2016). Transiently delivered scRNA/dCas9-TET1 induced demethylation of the targeted *RANKL* gene, a gene silenced through hypermethylation in HEK-293FT cells, and about 2-fold increased *RANKL* mRNA levels. Stable expression of the system yielded stronger effects:

10- to 20-fold transactivation (Xu *et al.*, 2016). To demonstrate the generic applicability of this epigenome editing strategy, expression of *MAGEB2* and *MMP2*, which are both silenced by hypermethylation in HeLa cells, was shown to be up-regulated by targeted demethylation. Interestingly, combining multiple sgRNAs that target distinct sites in the *RANKL* and *MAGE* genes does not result in significant additive or synergistic increase in transactivation (Xu *et al.*, 2016). A similar non-additive effect in transactivation was observed for dCas9-p300 targeted to multiple adjacent sites to induce histone acetylation (Hilton *et al.*, 2015).

Characterizing known or discovering new regulatory elements in the genome

Large-scale mapping of the human epigenome and chromatin states reveals vast numbers of putative regulatory elements (ENCODE Project Consortium, 2012; Thurman *et al.*, 2012; Roadmaps Epigenomics Consortium *et al.*, 2015). The validation and characterization of these elements benefits from the use of targetable epigenetic effectors to establish the functions of these regions in the genome.

In order to establish whether dCas9-LSD1 can be used to investigate enhancer function, the fusion protein has been targeted at well-characterized and putative enhancers in mouse embryonic stem cells. Before the introduction of dCas9, TALE–LSD1 fusion proteins had been already used to target endogenous candidate regulatory elements enriched for H3K4me2 and H3K27ac (Mendenhall *et al.*, 2013). These studies established that at a subset of putative enhancers the epigenetic state was altered following LSD1 targeting, permitting identification of the down-regulated target genes (Mendenhall *et al.*, 2013). Targeting of LSD1 to eight putative pluripotency-specific enhancers using dCas9 resulted in the discovery of four enhancers regulating the expression of genes critical in maintaining embryonic state (Kearns *et al.*, 2015). Chromosome conformation capture (3C) confirmed that the target gene was not down-regulated through enhancer–promoter loop disturbance of dCas9-LSD1 binding *per se*. A reduction in H3K4me2 and H3K27ac was observed around the sgRNA-dCas9-LSD1 target site in the identified *Tbx3*-enhancer. In contrast, dCas9-KRAB targeted to the same site in the *Tbx3*-enhancer resulted in a reduction of H3K27ac and an increase of H3K27me3 and H3K9me3 at the *Tbx3* proximal promoter region. However, neither dCas9-KRAB nor dCas9-LSD1 was able to increase the level of the repressive marks H3K27me3 and H3K9me3 at the *Tbx3*-enhancer region. In a different study, in which the KRAB domain was targeted at the HS2 enhancer in the globin locus control region (LCR) in human K562 cells, H3K9me3 marks were effectively introduced at the target enhancer region (Thakore *et al.*, 2015). The introduction of these repressive marks was concurrent with decreased chromatin accessibility at the target site in the enhancer as well as in several other parts of the LCR (Thakore *et al.*, 2015). These findings suggest pleiotropic effects of dCas9-KRAB at the targeted enhancer–promoter loop or as yet uncharacterized downstream mechanisms of chromatin reorganization.

Genome-wide forward screening

Using libraries of targeted DNA-binding platforms fused to effectors the epigenetic states and/or expression levels of genes can be altered to study their impact on selected downstream processes or phenotypes in genome-wide forward screens. Such screening applications gain momentum due to expanding number of high-resolution techniques to measure epigenetic landscapes and chromatin states (Zentner and Henikoff, 2014; Ramani *et al.*, 2016), including the introduction of novel methods for mapping epigenetic modifications in the genomes of single cells (Angermueller *et al.*, 2016; Song *et al.*, 2016).

The application of dCas9 in forward screening using targeted transcription factors (TTFs) was shown to permit highly specific genome-scale transcription modulation (Gilbert *et al.*, 2014; Konermann *et al.*, 2015). Using combinations of 10 sgRNAs per gene, tiling −50 to +300 bp for repression (CRISPRi) or −400 to −50 bp for activation (CRISPRa) around the TSS to target nearly 1600 protein-encoding genes, human K562 cells were screened for growth phenotypes. In this study transcription was repressed using dCas9-KRAB and activated using dCas9-SunTag recruiting scFv–VP64 fusions (Gilbert *et al.*, 2014). In a second screen, known as well as novel complexes and pathways involved in the response to a chimeric cholera–diphtheria toxin were identified (Gilbert *et al.*, 2014). Using next-generation sequencing, sgRNAs guiding the dCas9-effectors could afterwards be identified to determine the genomic locus responsible for the screened phenotypic characteristics (Gilbert *et al.*, 2014; Konermann *et al.*, 2015). A crucial aspect of screening studies using dCas9–effector fusions is that due to improved sgRNA design and more effective dCas9 design strategies, one dCas9 recruited by a single sgRNA now suffices in modulating a specific locus (Konermann *et al.*, 2015).

Current hurdles in application of dCas9

dCas9 has been shown to exhibit off-target binding at genome-wide scale (Kuscu *et al.*, 2014; Tsai *et al.*, 2015). Studies on target specificity have mainly been done on Cas9 (Fu *et al.*, 2014; Slaymaker *et al.*, 2016; Kleinstiver *et al.*, 2016); these findings can be extrapolated to dCas9 and dCas9-based effectors. The ability of the Cas9–sgRNA complex to localize and bind at a target sequence could be altered by genome accessibility as a consequence of local chromatin environment (Knight *et al.*, 2015) or binding of endogenous proteins such as transcription factors (Hilton *et al.*, 2015). Taking that notion into account target sites can be selected based on information regarding genome-bound proteins (e.g. using DNase-seq) (Maeder *et al.*, 2013b; Kearns *et al.*, 2014; Thakore *et al.*, 2015). Alternatively, DNA accessibility and off-target binding are determined *ad hoc* for every application. Off-target binding is related to flexibility in PAM sequence recognition and permitted mismatches in the PAM-distal part of the sgRNA (Kim *et al.*, 2015; Ran *et al.*, 2015; Tsai *et al.*, 2015; Wang *et al.*, 2015b; Leenay *et al.*, 2016). Efforts to reduce off-target binding include rational redesign of Cas9 for higher PAM specificity (Kleinstiver *et al.*, 2015), reduction of non-specific charge interactions between Cas9 and non-target DNA strand (Slaymaker *et al.*, 2016) or hydrogen bonding between Cas9 and the backbones of the target DNA strand and the sgRNA (Kleinstiver *et al.*, 2016), as well as use of truncated sgRNAs (Fu *et al.*, 2014). Although many of these strategies lead to a reduction of the DNA binding affinity of Cas9 or its activity, these efforts have contributed significantly to overcoming Cas9 off-target binding.

Another issue for applications of dCas9 is its size. The coding sequence of the widely used SpyCas9 is around 4.2 kb, which is compatible with the ~4.5 kb maximum accommodated in common viral vectors used for delivery in mammalian cells. However, such a long dCas9 coding sequence does not leave 'space' for sgRNA sequences and regulatory elements to be co-delivered. This limits the application of some more complex design strategies such as extending the sgRNA with aptamers, adding effector domains to dCas9 or using pools of multiple sgRNAs. The relatively small Cas9 orthologue from *Staphylococcus aureus* (SaCas9) is 3.2 kb in size (Friedland *et al.*, 2015; Ran *et al.*, 2015). The use of such small orthologues reduces the size problem with no drawbacks on activity, as observed for a truncated SpyCas9 derivative in which a less-conserved, non-interacting portion of the recognition lobe had been removed (Nishimasu *et al.*, 2014). The split-dCas9 system also provides a solution to the size problem, but it requires two vectors for delivery (Truong *et al.*, 2015). Alternatively, cells stably expressing dCas9-effector proteins can be complemented by delivery of sgRNA, but stable expression is difficult to achieve when using primary cells or multicellular organisms. Another option is dCas9 delivery using *in vitro* assembled sgRNA-(d)Cas9 ribonucleoproteins (RNPs), an approach which was successful for dCas9-VP64 (Zuris *et al.*, 2015) and Cas9 genome editing applications (Cho *et al.*, 2013b; Kim *et al.*, 2014; Ramakrishna *et al.*, 2014; Sun *et al.*, 2015; Zuris *et al.*, 2015; D'Astolfo *et al.*, 2015). Finally, a potential hurdle for implementation in living organisms is the fact that the (d) Cas9 protein elicits an immunogenic responses due to its exogenous origin (Wang *et al.*, 2015a). Future dedicated studies will need to clarify the extent of this issue for possible clinical applications.

Conclusions and future perspectives

The importance of the epigenome is becoming increasingly clear in the light of developmental processes, cell differentiation status and diseases. In an attempt to draw causal relations rather than correlations to understand and control the interplay between the epigenome, chromatin structure and gene expression levels, targetable DNA-binding proteins have shown to be of great value. dCas9 has a more robust yet readily adjustable DNA-recognition mechanism compared to ZFPs and TALEs; it has proven to be a flexible tool in recruiting effector domains to targeted DNA-sequences, in particular due to the numerous options arising from expanding the sgRNA even up to 5 kb long autonomous functional RNAs (Shechner *et al.*, 2015). The full value of dCas9 in the field of epigenetics has yet to be revealed with more functional studies at local scales and genome-wide level, providing detailed causal relations between epigenetic modifications, chromatin structure and gene expression. Future application in translational sciences such as medicine, synthetic biology or biotechnology, rely on advances in overcoming final hurdles provided by delivery problems due to its size, possible immunogenic response and off-target effects.

Alternatives for dCas9 emerge from the discovery and characterization of novel and known CRISPR systems involving protein complexes with different DNA-recognition mechanisms or different dimensions. Possible alternatives have recently become available in the form of the relatively small Cpf1 (~3.9 kb) (Zetsche *et al.*, 2015a; Yamano *et al.*, 2016), the C2c1 and C2c3 systems (Shmakov *et al.*, 2015) or even the multiprotein Cascade complex from the abundant Type I-E system native to *E. coli* K-12 (Luo *et al.*, 2015). These and future related systems add to a comprehensive CRISPR-based toolbox with distinct and orthogonal guiding crRNA, protospacer- or PAM-recognition, which is destined to occupy a permanent and prominent place in molecular biology.

Glossary

VP64: A strong transcriptional activation domain that recruits a variety of transcription factors and chromatin remodelling factors (Cress and Triezenberg, 1991). Tetrameric fusion of the acidic domain of herpes simplex viral protein 16 (VP16) (Beerli *et al.*, 1998).

p65: The 65 kDa subunit of the NF-κB transcription factor (Schmitz and Baeuerle, 1991). Often the principal transactivation domain (TA_1) is used as activation domain (p65AD).

KRAB: Krüppel-associated box (KRAB) is a common repression domain in eukaryotic transcription factors (Margolin *et al.*, 1994). It associates with KAP1, which forms a scaffold to recruit several proteins involved in inducing and spreading of heterochromatin over large distances (Groner *et al.*, 2010).

p300: Human transcriptional regulatory protein able to acetylate histone 3 at lysine residue 27 (Ogryzko *et al.*, 1996), a signal for active transcription. Often the isolated catalytic core of p300, containing the histone acetyltransferase (HAT) domain, is used as effector.

CBP: The CREB-binding protein (CBP) is able to bind many different transcription factors, acting as scaffold to co-activate transcription (Chan and La Thangue, 2001). Additionally, it has a histone acetyl transferase (HAT) domain, which can be used as epigenetic effector to acetylate lysine residues in histones.

LSD1: Lysine-specific demethylase 1 (LSD1) demethylates histone 3 at lysine residues 4 and 9, leading to silencing of enhancers (Shi *et al.*, 2004). LSD1 functionally interacts with other chromatin-modifying enzymes, including histone deacetylases that remove acetyl groups from histone 3 at lysine residue 27 (Lee *et al.*, 2006).

DNMT3A: DNA methyltransferase 3A (DNMT3A) catalyses *de novo* DNA methylation as well as methylation of hemimethylated DNA, preferentially at CpG sites (Okano *et al.*, 1998; Oka *et al.*, 2006). Methylated CpGs in promoters are associated with silenced transcription.

TET1: Ten-eleven translocation (TET) methylcytosine dioxygenase (TET1) is able to oxidate 5-methyl cytosines (5mC) to 5-hydroxymethylcytosines (5hmC), a step in the active removal of 5mC (Tahiliani *et al.*, 2009; Xu *et al.*, 2011). At promoters, this oxidation and removal is associated with activation of silenced genes.

Acknowledgements

Research on the topic of this review in the lab of R.T.D. is supported by a grant from the Netherlands Organization for Scientific Research (VICI 016.160.613), M.T.-D. is supported by a grant from the Dutch Technology Foundation STW. D.J.W.B. contributed to the writing of this review article as part of the MSc programme in Life Science and Technology at Leiden University.

References

Angermueller, C., Clark, S.J., Lee, H.J., Macaulay, I.C., Teng, M.J., Hu, T.X., Krueger, F., Smallwood, S.A., Ponting, C.P., Voet, T., *et al.* (2016). Parallel single-cell sequencing links transcriptional and epigenetic heterogeneity. Nat.

Methods *13*, 229–232. http://dx.doi.org/10.1038/nmeth.3728

Anton, T., Bultmann, S., Leonhardt, H., and Markaki, Y. (2014). Visualization of specific DNA sequences in living mouse embryonic stem cells with a programmable fluorescent CRISPR/Cas system. Nucleus *5*, 163–172. http://dx.doi.org/10.4161/nucl.28488

Banaszynski, L.A., Liu, C.W., and Wandless, T.J. (2005). Characterization of the FKBP.rapamycin.FRB ternary complex. JACS *127*, 4715–4721. http://dx.doi.org/10.1021/ja043277y

Barrangou, R., Fremaux, C., Deveau, H., Richards, M., Boyaval, P., Moineau, S., Romero, D.A., and Horvath, P. (2007). CRISPR provides acquired resistance against viruses in prokaryotes. Science *315*, 1709–1712.

Baylin, S.B., and Jones, P.A. (2011). A decade of exploring the cancer epigenome - biological and translational implications. Nat. Rev. Cancer. *11*, 726–734. http://dx.doi.org/10.1038/nrc3130

Beerli, R.R., Segal, D.J., Dreier, B., and Barbas III, C.F. (1998). Toward controlling gene expression at will: specific regulation of the erbB-2/HER-2 promoter by using polydactyl zinc finger proteins constructed from modular building blocks. Proc. Natl. Acad. Sci. U.S.A. *95*, 14628–14633.

Berger, S.L., Kouzarides, T., Shiekhattar, R., and Shilatifard, A. (2009). An operational definition of epigenetics. Genes. Dev. *23*, 781–783. http://dx.doi.org/10.1101/gad.1787609

Bernstein, B.E., Stamatoyannopoulos, J.A., Costello, J.F., Ren, B., Milosavljevic, A., Meissner, A., Kellis, M., Marra, M.A., Beaudet, A.L., Ecker, J.R., *et al.* (2010). The NIH Roadmap Epigenomics Mapping Consortium. Nat. Biotechnol. *28*, 1045–1048. http://dx.doi.org/10.1038/nbt1010-1045

Bhaya, D., Davison, M., and Barrangou, R. (2011). CRISPR-Cas systems in bacteria and archaea: versatile small RNAs for adaptive defense and regulation. Annu. Rev. Genet. *45*, 273–297. http://dx.doi.org/10.1146/annurev-genet-110410-132430

Bikard, D., Jiang, W., Samai, P., Hochschild, A., Zhang, F., and Marraffini, L.A. (2013). Programmable repression and activation of bacterial gene expression using an engineered CRISPR-Cas system. Nucleic Acids Res. *41*, 7429–7437. http://dx.doi.org/10.1093/nar/gkt520

Bird, A. (2002). DNA methylation patterns and epigenetic memory. Genes. Dev. *16*, 6–21. http://dx.doi.org/10.1101/gad.947102

Bird, A. (2007). Perceptions of epigenetics. Nature *447*, 396–398.

Boch, J., Scholze, H., Schornack, S., Landgraf, A., Hahn, S., Kay, S., Lahaye, T., Nickstadt, A., and Bonas, U. (2009). Breaking the code of DNA binding specificity of TAL-type III effectors. Science *326*, 1509–1512. http://dx.doi.org/10.1126/science.1178811

Bultmann, S., Morbitzer, R., Schmidt, C.S., Thanisch, K., Spada, F., Elsaesser, J., Lahaye, T., and Leonhardt, H. (2012). Targeted transcriptional activation of silent oct4 pluripotency gene by combining designer TALEs and inhibition of epigenetic modifiers. Nucleic Acids Res. *40*, 5368–5377. http://dx.doi.org/10.1093/nar/gks199

Chan, H.M., and La Thangue, N.B. (2001). p300/CBP proteins: HATs for transcriptional bridges and scaffolds. J. Cell Sci. *114*, 2363–2373.

Chavez, A., Scheiman, J., Vora, S., Pruitt, B.W., Tuttle, M., Iyer, E.P.R., Lin, S., Kiani, S., Guzman, C.D., Wiegand, D.J., *et al.* (2015). Highly efficient Cas9-mediated transcriptional programming. Nat. Methods *12*, 326–328. http://dx.doi.org/10.1038/nmeth.3312

Chen, B., Gilbert, L.A., Cimini, B.A., Schnitzbauer, J., Zhang, W., Li, G.-W., Park, J., Blackburn, E.H., Weissman, J.S., Qi, L.S., *et al.* (2013). Dynamic imaging of genomic loci in living human cells by an optimized CRISPR/Cas system. Cell *155*, 1479–1491. http://dx.doi.org/10.1016/j.cell.2013.12.001

Cheng, A.W., Jillette, N., Lee, P., Plaskon, D., Fujiwara, Y., Wang, W., Taghbalout, A., and Wang, H. (2016). Casilio: a versatile CRISPR-Cas9-Pumilio hybrid for gene regulation and genomic labeling. Cell Res. *26*, 254–257. http://dx.doi.org/10.1038/cr.2016.3

Cheng, A.W., Wang, H., Yang, H., Shi, L., Katz, Y., Theunissen, T.W., Rangarajan, S., Shivalila, C.S., Dadon, D.B., and Jaenisch, R. (2013). Multiplexed activation of endogenous genes by CRISPR-on, an RNA-guided transcriptional activator system. Cell Res. *23*, 1163–1171. http://dx.doi.org/10.1038/cr.2013.122

Cho, S.W., Kim, S., Kim, J.M., and Kim, J.-S. (2013a). Targeted genome engineering in human cells with the Cas9 RNA-guided endonuclease. Nat. Biotechnol. *31*, 230–232. http://dx.doi.org/10.1038/nbt.2507

Cho, S.W., Lee, J., Carroll, D., Kim, J.-S., and Lee, J. (2013b). Heritable gene knockout in Caenorhabditis elegans by direct injection of Cas9-sgRNA ribonucleoproteins. Genetics *195*, 1177–1180. http://dx.doi.org/10.1534/genetics.113.155853

Chong, S., and Whitelaw, E. (2004). Epigenetic germline inheritance. Curr. Opin. Genet. Dev. *14*, 692–696. http://dx.doi.org/10.1016/j.gde.2004.09.001

Choo, Y. (1998). Recognition of DNA methylation by zinc fingers. Nat. Struct. Biol. 5, 264–265.

Choo, Y., Sánchez-García, I., and Klug, A. (1994). In vivo repression by a site-specific DNA-binding protein designed against an oncogenic sequence. Nature *372*, 642–645. http://dx.doi.org/10.1038/372642a0

Chylinski, K., Makarova, K.S., Charpentier, E., and Koonin, E.V. (2014). Classification and evolution of type II CRISPR-Cas systems. Nucleic Acids Res. *42*, 6091–6105. http://dx.doi.org/10.1093/nar/gku241

Cong, L., Ran, F.A., Cox, D., Lin, S., Barretto, R., Habib, N., Hsu, P.D., Wu, X., Jiang, W., Marraffini, L.A., *et al.* (2013). Multiplex genome engineering using CRISPR/Cas systems. Science *339*, 819–823. http://dx.doi.org/10.1126/science.1231143

Cress, W.D., and Triezenberg, S.J. (1991). Critical structural elements of the VP16 transcriptional activation domain. Science *251*, 87–90.

Cubas, P., Vincent, C., and Coen, E. (1999). An epigenetic mutation responsible for natural variation in floral symmetry. Nature *401*, 157–161. http://dx.doi.org/10.1038/43657

D'Astolfo, D.S., Pagliero, R.J., Pras, A., Karthaus, W.R., Clevers, H., Prasad, V., Lebbink, R.J., Rehmann, H., and Geijsen, N. (2015). Efficient intracellular delivery of native proteins. Cell *161*, 674–690. http://dx.doi.org/10.1016/j.cell.2015.03.028

Daniel, J.M., Spring, C.M., Crawford, H.C., Reynolds, A.B., and Baig, A. (2002). The p120ctn-binding partner Kaiso is a bi-modal DNA-binding protein that recognizes both a sequence-specific consensus and methylated CpG dinucleotides. Nucleic Acids Res. *30*, 2911–2919.

Davis, K.M., Pattanayak, V., Thompson, D.B., Zuris, J.A., and Liu, D.R. (2015). Small molecule-triggered Cas9 protein with improved genome-editing specificity. Nat. Chem. Biol. *11*, 316–318. http://dx.doi.org/10.1038/nchembio.1793

de Groote, M.L., Verschure, P.J., and Rots, M.G. (2012). Epigenetic Editing: targeted rewriting of epigenetic marks to modulate expression of selected target genes. Nucleic Acids Res. *40*, 10596–10613. http://dx.doi.org/10.1093/nar/gks863

Delvecchio, M., Gaucher, J., Aguilar-Gurrieri, C., Ortega, E., and Panne, D. (2013). Structure of the p300 catalytic core and implications for chromatin targeting and HAT regulation. Nat. Struct. Mol. Biol. *20*, 1040–1046. http://dx.doi.org/10.1038/nsmb.2642

Deng, D., Yan, C., Pan, X., Mahfouz, M., Wang, J., Zhu, J.-K., Shi, Y., and Yan, N. (2012). Structural basis for sequence-specific recognition of DNA by TAL effectors. Science *335*, 720–723. http://dx.doi.org/10.1126/science.1215670

Doudna, J.A., and Charpentier, E. (2014). Genome editing. The new frontier of genome engineering with CRISPR-Cas9. Science *346*, 1258096. http://dx.doi.org/10.1126/science.1258096

Dow, L.E., Fisher, J., O'Rourke, K.P., Muley, A., Kastenhuber, E.R., Livshits, G., Tschaharganeh, D.F., Socci, N.D., and Lowe, S.W. (2015). Inducible in vivo genome editing with CRISPR-Cas9. Nat. Biotechnol. *33*, 390–394. http://dx.doi.org/10.1038/nbt.3155

ENCODE Project Consortium. (2012). An integrated encyclopedia of DNA elements in the human genome. Nature *489*, 57–74. http://dx.doi.org/10.1038/nature11247

Esvelt, K.M., Mali, P., Braff, J.L., Moosburner, M., Yaung, S.J., and Church, G.M. (2013). Orthogonal Cas9 proteins for RNA-guided gene regulation and editing. Nat. Methods *10*, 1116–1121. http://dx.doi.org/10.1038/nmeth.2681

Friedland, A.E., Baral, R., Singhal, P., Loveluck, K., Shen, S., Sanchez, M., Marco, E., Gotta, G.M., Maeder, M.L., Kennedy, E.M., *et al.* (2015). Characterization of Staphylococcus aureus Cas9: a smaller Cas9 for all-in-one adeno-associated virus delivery and paired nickase applications. Genome Biol. *16*. http://dx.doi.org/10.1186/s13059-015-0817-8

Fu, Y., Sander, J.D., Reyon, D., Cascio, V.M., and Joung, J.K. (2014). Improving CRISPR-Cas nuclease specificity using truncated guide RNAs. Nat. Biotechnol. *32*, 279–284. http://dx.doi.org/10.1038/nbt.2808

Gaj, T., Gersbach, C.A., and Barbas III, C.F. (2013). ZFN, TALEN, and CRISPR/Cas-based methods for genome engineering. Trends Biotechnol. *31*, 397–405. http://dx.doi.org/10.1016/j.tibtech.2013.04.004

Gardner, K.E., Allis, C.D., and Strahl, B.D. (2011). Operating on chromatin, a colorful language where context matters. J. Mol. Biol. *409*, 36–46. http://dx.doi.org/10.1016/j.jmb.2011.01.040

Gersbach, C.A., Gaj, T., and Barbas III, C.F. (2014). Synthetic zinc finger proteins: the advent of targeted gene regulation and genome modification technologies. Acc. Chem. Res. *47*, 2309–2318. http://dx.doi.org/10.1021/ar500039w

Gilbert, L.A., Horlbeck, M.A., Adamson, B., Villalta, J.E., Chen, Y., Whitehead, E.H., Guimaraes, C., Panning, B., Ploegh, H.L., Bassik, M.C., *et al.* (2014). Genome-scale CRISPR-mediated control of gene repression and activation. Cell *159*, 647–661. http://dx.doi.org/10.1016/j.cell.2014.09.029

Gilbert, L.A., Larson, M.H., Morsut, L., Liu, Z., Brar, G.A., Torres, S.E., Stern-Ginossar, N., Brandman, O., Whitehead, E.H., Doudna, J.A., *et al.* (2013). CRISPR-mediated modular RNA-guided regulation of transcription in eukaryotes. Cell *154*, 442–451. http://dx.doi.org/10.1016/j.cell.2013.06.044

González, F., Zhu, Z., Shi, Z.D., Lelli, K., Verma, N., Li, Q.V., and Huangfu, D. (2014). An iCRISPR platform for rapid, multiplexable, and inducible genome editing in human pluripotent stem cells. Cell Stem Cell *15*, 215–226. http://dx.doi.org/10.1016/j.stem.2014.05.018

Grau, J., Boch, J., and Posch, S. (2013). TALENoffer: genome-wide TALEN off-target prediction. Bioinformatics *29*, 2931–2932. http://dx.doi.org/10.1093/bioinformatics/btt501

Groner, A.C., Meylan, S., Ciuffi, A., Zangger, N., Ambrosini, G., Dénervaud, N., Bucher, P., and Trono, D. (2010). KRAB-zinc finger proteins and KAP1 can mediate long-range transcriptional repression through heterochromatin spreading. PLoS Genet. *6*, e1000869. http://dx.doi.org/10.1371/journal.pgen.1000869

Guilinger, J.P., Pattanayak, V., Reyon, D., Tsai, S.Q., Sander, J.D., Joung, J.K., and Liu, D.R. (2014). Broad specificity profiling of TALENs results in engineered nucleases with improved DNA-cleavage specificity. Nat. Methods *11*, 429–435. http://dx.doi.org/10.1038/nmeth.2845

Hemphill, J., Borchardt, E.K., Brown, K., Asokan, A., and Deiters, A. (2015). Optical control of CRISPR/Cas9 gene editing. JACS *137*, 5642–5645. http://dx.doi.org/10.1021/ja512664v

Henikoff, S., and Shilatifard, A. (2011). Histone modification: cause or cog? Trends Genet. *27*, 389–396. http://dx.doi.org/10.1016/j.tig.2011.06.006

Hilton, I.B., D'Ippolito, A.M., Vockley, C.M., Thakore, P.I., Crawford, G.E., Reddy, T.E., and Gersbach, C.A. (2015). Epigenome editing by a CRISPR-Cas9-based acetyltransferase activates genes from promoters and enhancers. Nat. Biotechnol. 33, 510–517. http://dx.doi.org/10.1038/nbt.3199

Hirano, H., Gootenberg, J.S., Horii, T., Abudayyeh, O.O., Kimura, M., Hsu, P.D., Nakane, T., Ishitani, R., Hatada, I., Zhang, F., *et al.* (2016). Structure and engineering of *Francisella novicida* Cas9. Cell *164*, 950–961. http://dx.doi.org/10.1016/j.cell.2016.01.039

Holliday, R. (1987). The inheritance of epigenetic defects. Science 238, 163–170.

Holliday, R. (2006). Epigenetics: a historical overview. Epigenetics *1*, 76–80. http://dx.doi.org/10.4161/epi.1.2.2762

Hwang, W.Y., Fu, Y., Reyon, D., Maeder, M.L., Tsai, S.Q., Sander, J.D., Peterson, R.T., Yeh, J.R., and Joung, J.K. (2013). Efficient genome editing in zebrafish using a CRISPR-Cas system. Nat. Biotechnol. *31*, 227–229. http://dx.doi.org/10.1038/nbt.2501

Jinek, M., Chylinski, K., Fonfara, I., Hauer, M., Doudna, J.A., and Charpentier, E. (2012). A programmable dual-RNA-guided DNA endonuclease in adaptive bacterial immunity. Science *337*, 816–821. http://dx.doi.org/10.1126/science.1225829

Jinek, M., East, A., Cheng, A., Lin, S., Ma, E., and Doudna, J. (2013). RNA-programmed genome editing in human cells. Elife *2*, e00471. http://dx.doi.org/10.7554/eLife.00471

Kawano, F., Suzuki, H., Furuya, A., and Sato, M. (2015). Engineered pairs of distinct photoswitches for optogenetic control of cellular proteins. Nat. Commun. *6*, 6256. http://dx.doi.org/10.1038/ncomms7256

Kearns, N.A., Genga, R.M.J., Enuameh, M.S., Garber, M., Wolfe, S.A., and Maehr, R. (2014). Cas9 effector-mediated regulation of transcription and differentiation in human pluripotent stem cells. Development *141*, 219–223. http://dx.doi.org/10.1242/dev.103341

Kearns, N.A., Pham, H., Tabak, B., Genga, R.M., Silverstein, N.J., Garber, M., and Maehr, R. (2015). Functional annotation of native enhancers with a Cas9-histone demethylase fusion. Nat. Methods *12*, 401–403. http://dx.doi.org/10.1038/nmeth.3325

Kelly, T.K., De Carvalho, D.D., and Jones, P.A. (2010). Epigenetic modifications as therapeutic targets. Nat. Biotechnol. *28*, 1069–1078. http://dx.doi.org/10.1038/nbt.1678

Kim, D., Bae, S., Park, J., Kim, E., Kim, S., Yu, H.R., Hwang, J., Kim, J.-I., and Kim, J.-S. (2015). Digenome-seq: genome-wide profiling of CRISPR-Cas9 off-target effects in human cells. Nat. Methods *12*, 237–243. http://dx.doi.org/10.1038/nmeth.3284

Kim, S., Kim, D., Cho, S.W., Kim, J., and Kim, J.-S. (2014). Highly efficient RNA-guided genome editing in human cells via delivery of purified Cas9 ribonucleoproteins. Genome Res. *24*, 1012–1019. http://dx.doi.org/10.1101/gr.171322.113

Kleinstiver, B.P., Pattanayak, V., Prew, M.S., Tsai, S.Q., Nguyen, N.T., Zheng, Z., and Joung, J.K. (2016). High-fidelity CRISPR-Cas9 nucleases with no detectable genome-wide off-target effects. Nature *529*, 490–495. http://dx.doi.org/10.1038/nature16526

Kleinstiver, B.P., Prew, M.S., Tsai, S.Q., Topkar, V.V., Nguyen, N.T., Zheng, Z., Gonzales, A.P.W., Li, Z., Peterson, R.T., Yeh, J.R., *et al.* (2015). Engineered CRISPR-Cas9 nucleases with altered PAM specificities. Nature *523*, 481–485. http://dx.doi.org/10.1038/nature14592

Klug, A. (2010). The discovery of zinc fingers and their applications in gene regulation and genome manipulation. Annu. Rev. Biochem. *79*, 213–231. http://dx.doi.org/10.1146/annurev-biochem-010909-095056

Knight, S.C., Xie, L., Deng, W., Guglielmi, B., Witkowsky, L.B., Bosanac, L., Zhang, E.T., El Beheiry, M., Masson, J.B., Dahan, M., *et al.* (2015). Dynamics of CRISPR-Cas9 genome interrogation in living cells. Science *350*, 823–826. http://dx.doi.org/10.1126/science.aac6572

Konermann, S., Brigham, M.D., Trevino, A.E., Hsu, P.D., Heidenreich, M., Cong, L., Platt, R.J., Scott, D.A., Church, G.M., and Zhang, F. (2013). Optical control of mammalian endogenous transcription and epigenetic states. Nature *500*, 472–476. http://dx.doi.org/10.1038/nature12466

Konermann, S., Brigham, M.D., Trevino, A.E., Joung, J., Abudayyeh, O.O., Barcena, C., Hsu, P.D., Habib, N., Gootenberg, J.S., Nishimasu, H., *et al.* (2015). Genome-scale transcriptional activation by an engineered CRISPR-Cas9 complex. Nature *517*, 583–588. http://dx.doi.org/10.1038/nature14136

Kouzarides, T. (2007). Chromatin modifications and their function. Cell *128*, 693–705.

Kuscu, C., Arslan, S., Singh, R., Thorpe, J., and Adli, M. (2014). Genome-wide analysis reveals characteristics of off-target sites bound by the Cas9 endonuclease. Nat. Biotechnol. *32*, 677–683. http://dx.doi.org/10.1038/nbt.2916

Lawhorn, I.E.B., Ferreira, J.P., and Wang, C.L. (2014). Evaluation of sgRNA target sites for CRISPR-mediated repression of TP53. PLoS One *9*, e113232. http://dx.doi.org/10.1371/journal.pone.0113232

Lee, M.G., Wynder, C., Bochar, D.A., Hakimi, M.-A., Cooch, N., and Shiekhattar, R. (2006). Functional interplay between histone demethylase and deacetylase enzymes. Mol. Cell. Biol. *26*, 6395–6402. http://dx.doi.org/10.1128/MCB.00723-06

Leenay, R.T., Maksimchuk, K.R., Slotkowski, R.A., Agrawal, R.N., Gomaa, A.A., Briner, A.E., Barrangou, R., and Beisel, C.L. (2016). Identifying and visualizing functional PAM diversity across CRISPR-Cas systems. Mol. Cell *62*, 137–147. http://dx.doi.org/10.1016/j.molcel.2016.02.031

Luo, M.L., Mullis, A.S., Leenay, R.T., and Beisel, C.L. (2015). Repurposing endogenous type I CRISPR-Cas systems for programmable gene repression. Nucleic Acids Res. *43*, 674–681. http://dx.doi.org/10.1093/nar/gku971

Lyko, F., and Brown, R. (2005). DNA methyltransferase inhibitors and the development of epigenetic cancer therapies. J. Natl. Cancer Inst. *97*, 1498–1506. http://dx.doi.org/10.1093/jnci/dji311

Ma, H., Naseri, A., Reyes-Gutierrez, P., Wolfe, S.A., Zhang, S., and Pederson, T. (2015). Multicolor CRISPR labeling of chromosomal loci in human cells. Proc. Natl. Acad. Sci. U.S.A. *112*, 3002–3007. http://dx.doi.org/10.1073/pnas.1420024112

Maeder, M.L., Angstman, J.F., Richardson, M.E., Linder, S.J., Cascio, V.M., Tsai, S.Q., Ho, Q.H., Sander, J.D., Reyon, D., Bernstein, B.E., *et al.* (2013a). Targeted DNA demethylation and activation of endogenous genes using programmable TALE-TET1 fusion proteins. Nat. Biotechnol. *31*, 1137–1142. http://dx.doi.org/10.1038/nbt.2726

Maeder, M.L., Linder, S.J., Cascio, V.M., Fu, Y., Ho, Q.H., and Joung, J.K. (2013b). CRISPR RNA-guided activation of endogenous human genes. Nat. Methods *10*, 977–979. http://dx.doi.org/10.1038/nmeth.2598

Mali, P., Aach, J., Stranges, P.B., Esvelt, K.M., Moosburner, M., Kosuri, S., Yang, L., and Church, G.M. (2013a). CAS9 transcriptional activators for target specificity screening and paired nickases for cooperative genome engineering. Nat. Biotechnol. *31*, 833–838. http://dx.doi.org/10.1038/nbt.2675

Mali, P., Yang, L., Esvelt, K.M., Aach, J., Guell, M., DiCarlo, J.E., Norville, J.E., and Church, G.M. (2013b). RNA-guided human genome engineering via Cas9. Science *339*, 823–826. http://dx.doi.org/10.1126/science.1232033

Margolin, J.F., Friedman, J.R., Meyer, W.K.-H., Vissing, H., Thiesen, H.-J., and Rauscher III, F.J. (1994). Krüppel-associated boxes are potent transcriptional repression domains. Proc. Natl. Acad. Sci. U.S.A. *91*, 4509–4513.

McDonald, J.I., Celik, H., Rois, L.E., Fishberger, G., Fowler, T., Rees, R., Kramer, A., Martens, A., Edwards, J.R., and Challen, G.A. (2016). Reprogrammable CRISPR/Cas9-based system for inducing site-specific DNA methylation. Biol. Open. *5*, 866–874. http://dx.doi.org/10.1242/bio.019067

Memedula, S., and Belmont, A.S. (2003). Sequential recruitment of HAT and SWI/SNF components to condensed chromatin by VP16. Curr. Biol. *13*, 241–246. http://dx.doi.org/10.1016/S0960-9822(03)00048-4

Mendenhall, E.M., Williamson, K.E., Reyon, D., Zou, J.Y., Ram, O., Joung, J.K., and Bernstein, B.E. (2013). Locus-specific editing of histone modifications at endogenous enhancers. Nat. Biotechnol. *31*, 1133–1136. http://dx.doi.org/10.1038/nbt.2701

Mittler, G., Stühler, T., Santolin, L., Uhlmann, T., Kremmer, E., Lottspeich, F., Berti, L., and Meisterernst, M. (2003). A novel docking site on Mediator is critical for activation by VP16 in mammalian cells. EMBO J. *22*, 6494–6504. http://dx.doi.org/10.1093/emboj/cdg619

Morbitzer, R., Römer, P., Boch, J., and Lahaye, T. (2010). Regulation of selected genome loci using de novo-engineered transcription activator-like effector (TALE)-type transcription factors. Proc. Natl. Acad. Sci. U.S.A. *107*, 21617–21622. http://dx.doi.org/10.1073/pnas.1013133107

Moscou, M.J., and Bogdanove, A.J. (2009). A simple cipher governs DNA recognition by TAL effectors. Science *326*, 1501. http://dx.doi.org/10.1126/science.1178817

Mussolino, C., and Cathomen, T. (2012). TALE nucleases: tailored genome engineering made easy. Curr. Opin. Biotechnol. *23*, 644–650. http://dx.doi.org/10.1016/j.copbio.2012.01.013

Nihongaki, Y., Kawano, F., Nakajima, T., and Sato, M. (2015a). Photoactivatable CRISPR-Cas9 for optogenetic genome editing. Nat. Biotechnol. *33*, 755–760. http://dx.doi.org/10.1038/nbt.3245

Nihongaki, Y., Yamamoto, S., Kawano, F., Suzuki, H., and Sato, M. (2015b). CRISPR-Cas9-based photoactivatable transcription system. Chem. Biol. *22*, 169–174. http://dx.doi.org/10.1016/j.chembiol.2014.12.011

Nishimasu, H., Ran, F.A., Hsu, P.D., Konermann, S., Shehata, S.I., Dohmae, N., Ishitani, R., Zhang, F., and Nureki, O. (2014). Crystal structure of Cas9 in complex with guide RNA and target DNA. Cell *156*, 935–949. http://dx.doi.org/10.1016/j.cell.2014.02.001

Ogryzko, V.V., Schiltz, R.L., Russanova, V., Howard, B.H., and Nakatani, Y. (1996). The transcriptional coactivators p300 and CBP are histone acetyltransferases. Cell *87*, 953–959.

Oka, M., Rodić, N., Graddy, J., Chang, L.-J., and Terada, N. (2006). CpG sites preferentially methylated by Dnmt3a in vivo. J. Biol. Chem. *281*, 9901–9908. http://dx.doi.org/10.1074/jbc.M511100200

Okano, M., Xie, S., and Li, E. (1998). Cloning and characterization of a family of novel mammalian DNA (cytosine-5) methyltransferases. Nat. Genet. *19*, 219–220. http://dx.doi.org/10.1038/890

Pavletich, N.P., and Pabo, C.O. (1991). Zinc finger-DNA recognition: crystal structure of a Zif268-DNA complex at 2.1 Å. Science *252*, 809–817. http://dx.doi.org/10.1126/science.2028256

Perez-Pinera, P., Kocak, D.D., Vockley, C.M., Adler, A.F., Kabadi, A.M., Polstein, L.R., Thakore, P.I., Glass, K.A., Ousterout, D.G., Leong, K.W., *et al.* (2013). RNA-guided gene activation by CRISPR-Cas9-based transcription factors. Nat. Methods *10*, 973–976. http://dx.doi.org/10.1038/nmeth.2600

Plass, C., Pfister, S.M., Lindroth, A.M., Bogatyrova, O., Claus, R., and Lichter, P. (2013). Mutations in regulators of the epigenome and their connections to global chromatin patterns in cancer. Nat. Rev. Genet. *14*, 765–780. http://dx.doi.org/10.1038/nrg3554

Polstein, L.R., and Gersbach, C.A. (2015). A light-inducible CRISPR-Cas9 system for control of endogenous gene activation. Nat. Chem. Biol. *11*, 198–200. http://dx.doi.org/10.1038/nchembio.1753

Qi, L.S., Larson, M.H., Gilbert, L.A., Doudna, J.A., Weissman, J.S., Arkin, A.P., and Lim, W.A. (2013). Repurposing CRISPR as an RNA-guided platform for sequence-specific control of gene expression. Cell *152*, 1173–1183. http://dx.doi.org/10.1016/j.cell.2013.02.022

Ramakrishna, S., Abu-Bonsrah, K.D., Beloor, J., Gopalappa, R., Lee, S.-K., and Kim, H. (2014). Gene disruption by cell-penetrating peptide-mediated delivery of Cas9 protein and guide RNA. Genome Res. *24*, 1020–1027. http://dx.doi.org/10.1101/gr.171264.113

Ramani, V., Shendure, J., and Duan, Z. (2016). Understanding spatial genome organization: methods and insights. Genomics Proteomics Bioinformatics *14*, 7–20. http://dx.doi.org/10.1016/j.gpb.2016.01.002

Ran, F.A., Cong, L., Yan, W.X., Scott, D.A., Gootenberg, J.S., Kriz, A.J., Zetsche, B., Shalem, O., Wu, X., Makarova, K.S., *et al.* (2015). In vivo genome editing using *Staphylococcus aureus* Cas9. Nature *520*, 186–191. http://dx.doi.org/10.1038/nature14299

Reyon, D., Tsai, S.Q., Khayter, C., Foden, J.A., Sander, J.D., and Joung, J.K. (2012). FLASH assembly of TALENs for high-throughput genome editing. Nat. Biotechnol. *30*, 460–465. http://dx.doi.org/10.1038/nbt.2170

Riggsbee, C.W., and Deiters, A. (2010). Recent advances in the photochemical control of protein function. Trends Biotechnol. *28*, 468–475. http://dx.doi.org/10.1016/j.tibtech.2010.06.001

Rivera, C.M., and Ren, B. (2013). Mapping human epigenomes. Cell *155*, 39–55. http://dx.doi.org/10.1016/j.cell.2013.09.011

Roadmaps Epigenomics Consortium, Kundaje, A., Meuleman, W., Ernst, J., Bilenky, M., Yen, A., Heravi-Moussavi, A., Kheradpour, P., Zhang, Z., Wang, J., *et al.* (2015). Integrative analysis of 111 reference human epigenomes. Nature *518*, 317–330. http://dx.doi.org/10.1038/nature14248

Rogers, J.M., Barrera, L.A., Reyon, D., Sander, J.D., Kellis, M., Joung, J.K., and Bulyk, M.L. (2015). Context influences on TALE-DNA binding revealed by quantitative profiling. Nat. Commun. *6*, 7440. http://dx.doi.org/10.1038/ncomms8440

Sanjana, N.E., Cong, L., Zhou, Y., Cunniff, M.M., Feng, G., and Zhang, F. (2012). A transcription activator-like effector toolbox for genome engineering. Nat. Protoc. *7*, 171–192. http://dx.doi.org/10.1038/nprot.2011.431

Schmitz, M.L., and Baeuerle, P.A. (1991). The p65 subunit is responsible for the strong transcription activating potential of NF-kappaB. EMBO J. *10*, 3805–3817.

Schübeler, D. (2015). Function and information content of DNA methylation. Nature *517*, 321–326. http://dx.doi.org/10.1038/nature14192

Shechner, D.M., Hacisuleyman, E., Younger, S.T., and Rinn, J.L. (2015). Multiplexable, locus-specific targeting of long RNAs with CRISPR-Display. Nat. Methods *12*, 664–670. http://dx.doi.org/10.1038/nmeth.3433

Shi, Y., Lan, F., Matson, C., Mulligan, P., Whetstine, J.R., Cole, P.A., Casero, R.A., and Shi, Y. (2004). Histone demethylation mediated by the nuclear amine oxidase homolog LSD1. Cell *119*, 941–953. http://dx.doi.org/10.1016/j.cell.2004.12.012

Shmakov, S., Abudayyeh, O.O., Makarova, K.S., Wolf, Y.I., Gootenberg, J.S., Semenova, E., Minakhin, L., Joung, J., Konermann, S., Severinov, K., *et al.* (2015). Discovery and functional characterization of diverse Class 2 CRISPR-Cas SYSTEMS. Mol. Cell *60*, 385–397. http://dx.doi.org/10.1016/j.molcel.2015.10.008

Slaymaker, I.M., Gao, L., Zetsche, B., Scott, D.A., Yan, W.X., and Zhang, F. (2016). Rationally engineered Cas9 nucleases with improved specificity. Science *351*, 84–88. http://dx.doi.org/10.1126/science.aad5227

Smith, Z.D., and Meissner, A. (2013). DNA methylation: roles in mammalian development. Nat. Rev. Genet. *14*, 204–220. http://dx.doi.org/10.1038/nrg3354

Snowden, A.W., Gregory, P.D., Case, C.C., and Pabo, C.O. (2002). Gene-specific targeting of H3K9 methylation is sufficient for initiating repression in vivo. Curr. Biol. *12*, 2159–2166. http://dx.doi.org/10.1016/S0960-9822(02)01391-X

Song, C.-X., Diao, J., Brunger, A.T., and Quake, S.R. (2016). Simultaneous single-molecule epigenetic imaging of DNA methylation and hydroxymethylation. Proc. Natl. Acad. Sci. U.S.A. *113*, 4338–4343. http://dx.doi.org/10.1073/pnas.1600223113

Stancheva, I., and Meehan, R.R. (2000). Transient depletion of xDnmt1 leads to premature gene activation in *Xenopus* embryos. Genes Dev. *14*, 313–327.

Sun, W., Ji, W., Hall, J.M., Hu, Q., Wang, C., Beisel, C.L., and Gu, Z. (2015). Self-assembled DNA nanoclews for the efficient delivery of CRISPR-Cas9 for genome editing. Angew. Chem. Int. Ed. Engl. *54*, 12029–12033. http://dx.doi.org/10.1002/anie.201506030

Tahiliani, M., Koh, K.P., Shen, Y., Pastor, W.A., Bandukwala, H., Brudno, Y., Agarwal, S., Iyer, L.M., Liu, D.R., Aravind, L., *et al.* (2009). Conversion of 5-methylcytosine to 5-hydroxymethylcytosine in mammalian DNA by MLL partner TET1. Science *324*, 930–935. http://dx.doi.org/10.1126/science.1170116

Tanenbaum, M.E., Gilbert, L.A., Qi, L.S., Weissman, J.S., and Vale, R.D. (2014). A protein-tagging system for signal amplification in gene expression and fluorescence imaging. Cell *159*, 635–646. http://dx.doi.org/10.1016/j.cell.2014.09.039

Thakore, P.I., Black, J.B., Hilton, I.B., and Gersbach, C.A. (2016). Editing the epigenome: technologies for programmable transcription and epigenetic modulation. Nat. Methods *13*, 127–137. http://dx.doi.org/10.1038/nmeth.3733

Thakore, P.I., D'Ippolito, A.M., Song, L., Safi, A., Shivakumar, N.K., Kabadi, A.M., Reddy, T.E., Crawford, G.E., and Gersbach, C.A. (2015). Highly specific epigenome editing by CRISPR-Cas9 repressors for silencing of distal regulatory elements. Nat. Methods *12*, 1143–1149. http://dx.doi.org/10.1038/nmeth.3630

Thurman, R.E., Rynes, E., Humbert, R., Vierstra, J., Maurano, M.T., Haugen, E., Sheffield, N.C., Stergachis, A.B., Wang, H., Vernot, B., *et al.* (2012). The accessible chromatin landscape of the human genome. Nature *489*, 75–82. http://dx.doi.org/10.1038/nature11232

Truong, D.-J., Kühner, K., Kühn, R., Werfel, S., Engelhardt, S., Wurst, W., and Ortiz, O. (2015). Development of an intein-mediated split-Cas9 system for gene therapy. Nucleic Acids Res. *43*, 6450–6458. http://dx.doi.org/10.1093/nar/gkv601

Tsai, S.Q., Zheng, Z., Nguyen, N.T., Liebers, M., Topkar, V.V., Thapar, V., Wyvekens, N., Khayter, C., Iafrate, A.J., Le, L.P., *et al.* (2015). GUIDE-seq enables genome-wide profiling of off-target cleavage by CRISPR-Cas nucleases. Nat. Biotechnol. *33*, 187–197. http://dx.doi.org/10.1038/nbt.3117

Turner, B.M. (2012). The adjustable nucleosome: an epigenetic signaling module. Trends Genet. *28*, 436–444. http://dx.doi.org/10.1016/j.tig.2012.04.003

Urnov, F.D., Rebar, E.J., Holmes, M.C., Zhang, H.S., and Gregory, P.D. (2010). Genome editing with engineered zinc finger nucleases. Nat. Rev. Genet. *11*, 636–646. http://dx.doi.org/10.1038/nrg2842

van Essen, D., Engist, B., Natoli, G., and Saccani, S. (2009). Two modes of transcriptional activation at native promoters by NF-kappaB p65. PLoS Biol. *7*, e73. http://dx.doi.org/10.1371/journal.pbio.1000073

Vandevenne, M., Jacques, D.A., Artuz, C., Nguyen, C.D., Kwan, A.H.Y., Segal, D.J., Matthews, J.M., Crossley, M., Guss, J.M., and Mackay, J.P. (2013). New insights into DNA recognition by zinc fingers revealed by structural analysis of the oncoprotein ZNF217. J. Biol. Chem. *288*, 10616–10627. http://dx.doi.org/10.1074/jbc.M112.441451

Vardimon, L., Kressmann, A., Cedar, H., Maechler, M., and Doerfler, W. (1982). Expression of a cloned adenovirus gene is inhibited by in vitro methylation. Proc. Natl. Acad. Sci. U.S.A. *79*, 1073–1077.

Vojta, A., Dobrinić, P., Tadić, V., Bočkor, L., Korać, P., Julg, B., Klasić, M., and Zoldoš, V. (2016). Repurposing the CRISPR-Cas9 system for targeted DNA methylation. Nucleic Acids Res. *44*, 5615–5628. http://dx.doi.org/10.1093/nar/gkw159

Wang, D., Mou, H., Li, S., Li, Y., Hough, S., Tran, K., Li, J., Yin, H., Anderson, D.G., Sontheimer, E.J., *et al.* (2015a). Adenovirus-mediated somatic genome editing of Pten by CRISPR/Cas9 in mouse liver in spite of Cas9-specific immune responses. Hum. Gene Ther. *26*, 432–442. http://dx.doi.org/10.1089/hum.2015.087

Wang, L., Grossman, S.R., and Kieff, E. (2000). Epstein–Barr virus nuclear protein 2 interacts with p300, CBP, and PCAF histone acetyltransferases in activation of the LMP1 promoter. Proc. Natl. Acad. Sci. U.S.A. *97*, 430–435.

Wang, T., Wei, J.J., Sabatini, D.M., and Lander, E.S. (2014). Genetic screens in human cells using the CRISPR-Cas9 system. Science *343*, 80–84. http://dx.doi.org/10.1126/science.1246981

Wang, X., Wang, Y., Wu, X., Wang, J., Wang, Y., Qiu, Z., Chang, T., Huang, H., Lin, R.-J., and Yee, J.-K. (2015b). Unbiased detection of off-target cleavage by CRISPR-Cas9 and TALENs using integrase-defective lentiviral vectors. Nat. Biotechnol. *33*, 175–178. http://dx.doi.org/10.1038/nbt.3127

Webster, K.E., O'Bryan, M.K., Fletcher, S., Crewther, P.E., Aapola, U., Craig, J., Harrison, D.K., Aung, H., Phutikanit, N., Lyle, R., *et al.* (2005). Meiotic and epigenetic defects in Dnmt3L-knockout mouse spermatogenesis. Proc. Natl. Acad. Sci. U.S.A. *102*, 4068–4073. http://dx.doi.org/10.1073/pnas.0500702102

Wolfe, S.A., Grant, R.A., Elrod-Erickson, M., and Pabo, C.O. (2001). Beyond the "recognition code": structures of two Cys2His2 zinc finger/TATA box complexes. Structure *9*, 717–723. http://dx.doi.org/10.1016/S0969-2126(01)00632-3

Wolfe, S.A., Nekludova, L., and Pabo, C.O. (2000). DNA recognition by Cys2His2 zinc finger proteins. Annu. Rev. Biophys. Biomol. Struct. *29*, 183–212. http://dx.doi.org/10.1146/annurev.biophys.29.1.183

Wright, A.V., Sternberg, S.H., Taylor, D.W., Staahl, B.T., Bardales, J.A., Kornfeld, J.E., and Doudna, J.A. (2015). Rational design of a split-Cas9 enzyme complex. Proc. Natl. Acad. Sci. U.S.A. *112*, 2984–2989. http://dx.doi.org/10.1073/pnas.1501698112

Xu, X., Tao, Y., Gao, X., Zhang, L., Li, X., Zou, W., Ruan, K., Wang, F., Xu, G.-L., and Hu, R. (2016). A CRISPR-based approach for targeted DNA demethylation. Cell Discov. *2*, 16009. http://dx.doi.org/10.1038/celldisc.2016.9

Xu, Y., Wu, F., Tan, L., Kong, L., Xiong, L., Deng, J., Barbera, A.J., Zheng, L., Zhang, H., Huang, S., *et al.* (2011). Genome-wide regulation of 5hmC, 5mC, and gene expression by Tet1 hydroxylase in mouse embryonic stem cells. Mol. Cell *42*, 451–464. http://dx.doi.org/10.1016/j.molcel.2011.04.005

Yamano, T., Nishimasu, H., Zetsche, B., Hirano, H., Slaymaker, I.M., Li, Y., Fedorova, I., Nakane, T., Makarova, K.S., Koonin, E.V., *et al.* (2016). Crystal structure of Cpf1 in complex with guide RNA and target DNA. Cell *165*, 949–962. http://dx.doi.org/10.1016/j.cell.2016.04.003

Youngson, N.A., and Whitelaw, E. (2008). Transgenerational epigenetic effects. Annu. Rev. Genomics. Hum. Genet. *9*, 233–257. http://dx.doi.org/10.1146/annurev.genom.9.081307.164445

Zalatan, J.G., Lee, M.E., Almeida, R., Gilbert, L.A., Whitehead, E.H., La Russa, M., Tsai, J.C., Weissman, J.S., Dueber, J.E., Qi, L.S., *et al.* (2015). Engineering complex synthetic transcriptional programs with CRISPR RNA scaffolds. Cell *160*, 339–350. http://dx.doi.org/10.1016/j.cell.2014.11.052

Zentner, G.E., and Henikoff, S. (2014). High-resolution digital profiling of the epigenome. Nat. Rev. Genet. *15*, 814–827. http://dx.doi.org/10.1038/nrg3798

Zetsche, B., Gootenberg, J.S., Abudayyeh, O.O., Slaymaker, I.M., Makarova, K.S., Essletzbichler, P., Volz, S.E., Joung, J., van der Oost, J., Regev, A., *et al.* (2015a). Cpf1 is a single RNA-guided endonuclease of a class 2 CRISPR-Cas system. Cell *163*, 759–771. http://dx.doi.org/10.1016/j.cell.2015.09.038

Zetsche, B., Volz, S.E., and Zhang, F. (2015b). A split-Cas9 architecture for inducible genome editing and transcription modulation. Nat. Biotechnol. 33, 139–142. http://dx.doi.org/10.1038/nbt.3149

Zhang, F., Cong, L., Lodato, S., Kosuri, S., Church, G.M., and Arlotta, P. (2011). Efficient construction of sequence-specific TAL effectors for modulating mammalian transcription. Nat. Biotechnol. *29*, 149–154. http://dx.doi.org/10.1038/nbt.1775

Zhou, V.W., Goren, A., and Bernstein, B.E. (2011). Charting histone modifications and the functional organization of mammalian genomes. Nat. Rev. Genet. *12*, 7–18. http://dx.doi.org/10.1038/nrg2905

Zuris, J.A., Thompson, D.B., Shu, Y., Guilinger, J.P., Bessen, J.L., Hu, J.H., Maeder, M.L., Joung, J.K., Chen, Z.-Y., and Liu, D.R. (2015). Cationic lipid-mediated delivery of proteins enables efficient protein-based genome editing in vitro and in vivo. Nat. Biotechnol. 33, 73–80. http://dx.doi.org/10.1038/nbt.3081

Treating Genetic Disorders Using State-of-the-Art Technology

3

Muhammad Jamal[1]*, Arif Ullah[2], Muhammad Ahsan Naeem[1], Rohit Tyagi[1], Zeshan Habib[1], Faheem Ahmad Khan[3] and Khaista Rehman[1]

[1]State Key Laboratory of Agricultural Microbiology, College of Animal Science and Medicine, Huazhong Agricultural University, Wuhan, The People's Republic of China.
[2]Institute of Biochemistry and Biotechnology, University of Veterinary and Animal Sciences Lahore, Pakistan.
[3]Key Laboratory of Agricultural Animal Genetics, Breeding and Reproduction of the Ministry of Education, College of Animal Science and Technology, Huazhong Agricultural University, Wuhan, The People's Republic of China.

*Correspondence: jamalbiotech@webmail.hzau.edu.cn

https://doi.org/10.21775/9781910190630.03

Abstract
CRISPR/Cas9 [clustered regularly interspaced short palindromic repeats (CRISPR)-associated protein 9], basically a bacterial immune system, is now widely applicable to engineer genomes of a number of cells and organisms because of its simplicity and robustness. In research avenue the system has been optimized to regulate gene expression, modify epigenome and edit target locus. These applications make CRISPR/Cas9 a technology of choice to edit disease causing mutations as well as the epigenome more efficiently than ever before. Meanwhile its application in *in vivo* and *ex vivo* cells is encouraging the scientific community for more vigorous gene therapy and in clinical setups for therapeutic genome editing. Here we review the recent advances that CRISPR/Cas9 mediated genome editing has achieved and is reported in previous studies and address the challenges associated with it.

Introduction
Genome editing has a profound consequence on disease treatment. In order to understand the function of gene in the disease pathway, the modulation of its expression remains the basic aim of classical genetics and modern molecular biology (Barrangou *et al.*, 2015). The development of RNA interference technology in the early 1990s and its application in mammalian cells to unveil the molecular functions of genes gave rise to the era of reverse genetics. The research area was further strengthened by the development of genome editing tools based on DNA-binding nucleases such as zinc finger nucleases (ZFNs) and transcription activator-like nucleases (TALENs), and the recently discovered CRISPR/Cas9 system (Joung and Sander, 2013; Urnov *et al.*, 2010). The engineering of DNA binding proteins to target specific DNA sequences is time consuming and expensive. However, the simple design and high efficiency of CRISPR/Cas9 made it a famous genome editing tool, enabling genome editing in a number of organisms (Cho *et al.*, 2013; Ding *et al.*, 2013; Niu *et al.*, 2014; Wang *et al.*, 2013).

CRISPR/Cas9
CRISPR (clustered, regularly interspaced, short palindromic repeat)/Cas (CRISPR-associated protein) is an array of repeat and spacer sequences that work with Cas9 to protect bacteria and archaea against the invading genome (Wiedenheft *et al.*, 2012). About 90% of archaea and 40% of the sequenced bacterial genomes contain a CRISPR

system (Deveau *et al.*, 2008; Koonin and Makarova, 2009). Initially the system was identified by a Japanese research group in 1987; these repeat sequences were present in downstream of Iap gene (Ishino *et al.*, 1987). There are five different classes of CRISPR system among which Type II-A CRISPR system of *Streptococcus pyogenes* is well elaborated for its genome editing purpose. Naturally present Type II CRISPR/Cas9 system in *S. pyogenes* comprises a repeat-spacer-repeat sequence, which is transcribed into crRNA (CRISPR-RNA), the crRNA hybridizes with tracrRNA (trans-activating CrRNA) encoded by set of nucleotides present near the CRISPR locus. The hybridization of crRNA-tracrRNA known as guide RNA makes a ribonucleoprotein complex with Cas9 for invading DNA degradation (Doudna and Charpentier, 2014). In 2012, researchers used CRISPR/Cas9 from *S. pyogenes* for genome editing purposes (Jinek *et al.*, 2012). Cas9, after cleavage of double stranded DNA (dsDNA), activates the cell repair pathways to repair the dsDNA. These pathways include the erroneous pathways known as NHEJ (non-homologous end joining) which causes insertion or deletion mutations. In the presence of donor DNA (HDR template) the dsDNA is repaired by process known as HDR (homology directed repair) which causes knock-in/-out of gene of interest (Barnes, 2001; Dudás and Chovanec, 2004).

CRISPR/Cas9 has been applied both *in vivo* and *ex vivo* to treat the cells in individuals suffering from various genetic disorders, also optimized to generate disease model organisms (Table 3.1). To demonstrate the proof-of-concept for CRISPR/Cas9 as a therapeutic toolkit we summarize the ground breaking approaches carried out *ex vivo* and *in vivo* to treat the genetic flaws. Regarding genome editing, the delivery of Cas9–gRNA complex into the target organ or organ system can use different methods such as lipid mediated delivery, electroporation, cell penetrating peptides and purified Cas9 (Fig. 3.1). The *ex vivo* genome editing involves targeting the cells (somatic cells or progenitor cells) outside the body in culture followed by its reincorporation into the human body (Fig. 3.2).

Ex vivo genome editing

Zygote editing

The delivery of Cas9 (cassette, mRNA or purified protein), targeting guide sequence and HDR template (as a ssDNA or dsDNA) into the zygote or embryo can be successfully applied to edit any locus of interest in cells (Wang *et al.*, 2013). Given the permanent nature of such manipulation, the coming generation will acquire these changes possibly leading to the eradication of the disease. The

Table 3.1 List of genetic disorders treated or that can be treated with CRISPR/Cas9

Genetic disorder	Mutation	Target edited/or can be edited with CRISPR/Cas9	References
Cataracts	Multiple	*Crygc*	Wu *et al.*, 2013
β-Thalassaemia	Multiple	*HBB*	Liang *et al.*, 2015
Tyrosinaemia	Multiple	*Fah*	Yin *et al.*, 2014
Acute myeloid leukaemia	Multiple	*MLL3*	Chen *et al.*, 2014
High cholesterol level		*PCSK9*	Ding *et al.*, 2014
Fanconi anaemia	c.456+4A>T	*FANCC*	Osborn *et al.*, 2015
Myeloproliferative neoplasm polycythaemia vera	V617F point mutation	*JAK2*	Smith *et al.*, 2015
Cystic fibrosis	Codon deletion (encoding phenylalanine)	*CFTR*	Schwank *et al.*, 2013
Retinitis pigmentosa	Multiple	*RPGR*	Bassuk *et al.*, 2016
Corneal dystrophy	Multiple	*TGFBI*	Usui, 2016
Duchenne muscular dystrophy	Faulty exon	*DMD*	Long *et al.*, 2016
Sickle cell diseases	(A to T) in the 6th codon	*HBB*	Huang *et al.*, 2015
Polycythaemia vera	V61F	*JAK2*	Smith *et al.*, 2015

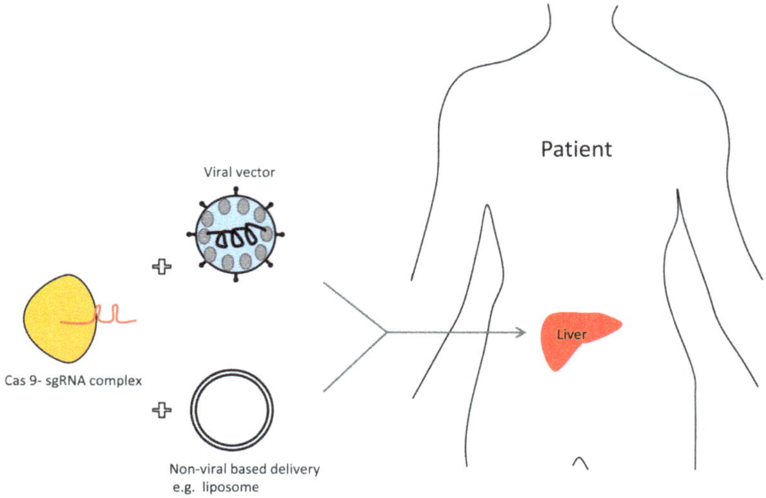

Figure 3.1 The CRISPR/Cas9 and sgRNA complex is packaged into virus or non-viral such as liquid nanoparticles to deliver it into target site, for example liver, or delivered systematically.

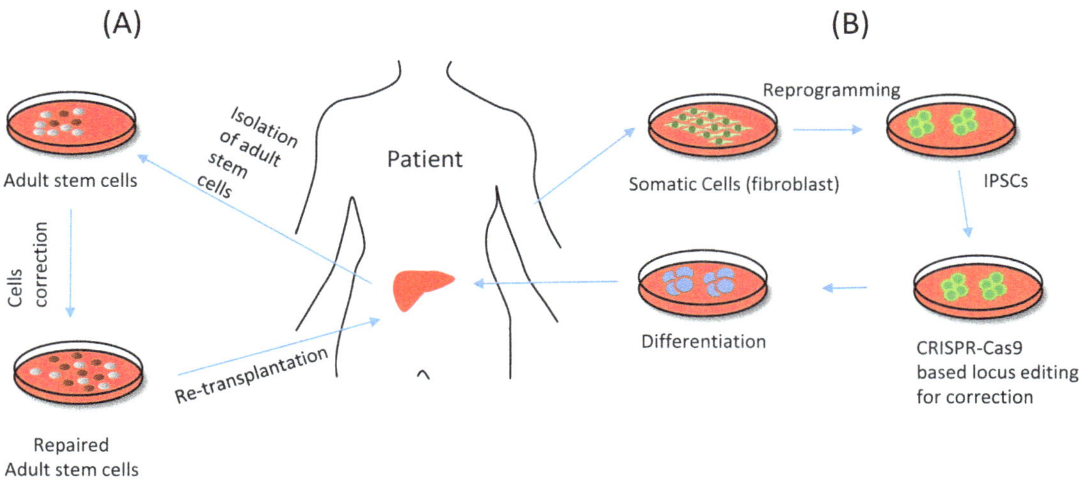

Figure 3.2 Schema of *ex vivo* genome editing. (A) adult stem cells such as haematopoietic stem cells (HSCs) and induced pluripotent stem cells (iPSCs) can be targeted for treatment followed by its reintroduction into the patient's body after clonal selection. (B) Another way is to isolate somatic cells from skin or blood such as a fibroblast, reprogramming into iPSCS, CRISPR/Cas9 mediated correction and its differentiation into HSCs followed by retransplantation into the patient's body.

first aforesaid application of CRISPR/dCas9 as a genome editing tool to treat genetic disorder was carried out by Wu *et al.* (2013) to treat an eye disorder known as cataracts. The disease is characterized by impairment in vision because of opaqueness in the eye lens. The gene responsible for this disease is *Crygc*, and in this study the researchers provided *in vitro*-transcribed Cas9 mRNA, gRNA and a template DNA containing the corrected gene of *Crygc* gene into the zygote.

In another approach the scientist treated Duchenne muscular dystrophy (DMD) by correcting the *dystrophin* gene. They applied Cas9 mRNA, gRNA and HDR template (as a ssDNA) into a mouse embryo to correct the muscular dystrophy (Long *et al.*, 2014).

In April 2015, a research group from China used the same system in non-viable human zygote to edit the β-globin gene *HBB* causing β-thalassaemia (Liang *et al.*, 2015), which created a huge controversy regarding the ethical concern in the scientific community. The scientist used the triponuclear zygote (egg fertilized by two sperms and having an extra set of chromosomes). Although they failed to achieve full-length efficiency and on-target activity, they are hopeful to optimize the system, especially for on-target specificity prior to clinical application.

The paternal genome editing allows the efficient integration of transgene into the genome compared with oocyte at sites know as quasi-random sites (Perry *et al.*, 2001; Yanagimachi, 2002). This is because, during fertilization, the decondensation (loss of nucleoprotein) makes it a better substrate for recombination. To this end, the injection of Cas9-sgRNA in decondensation phase followed by ICSI of unfertilized oocyte efficiently achieved editing of both *eGFP* and *Tyr* (Suzuki *et al.*, 2014).

In vivo genome editing

Remediation of haematological disorders

Tyrosinaemia, characterized by an elevated level of tyrosine in the blood, results in various disorders such as hereditary infantile tyrosinaemia (tyrosinaemia-I), tyrosinaemia-II and tyrosinaemia of the newborn (TTN). Tyrosinaemia-1 is also called Fanconi syndrome, with renal tubular failure in early stage of life in infants. The mutant *Fah* (fumarylacetoacetate hydrolase) gene responsible for the disease phenotype was corrected in an adult mouse by hydrodynamic delivery of a plasmid expressing Cas9 and sgRNA encoded in a plasmid rescued wild-type expression of Fah protein in 1 out of 250 cells (Yin *et al.*, 2014).

Acute myeloid leukaemia (AML) is cancer of the myeloid line of blood cells affecting adults. The incidence of disease increases with age because of deleted regions located on 7q on chromosomes. *MLL3* gene was identified as haploinsufficient tumour suppressor in AML. CRISPR/Cas9 was used to disrupt *MLL3* in $p53^{-/-}$ mouse haematopoietic stem/progenitor cells (HSPCs) which showed disease acceleration and AML development (Chen *et al.*, 2014). More recently, researchers from Wellcome Trust identified a potential gene known as *KAT2A*. CRISPR/Cas9-mediated gene disruption of this gene resulted in slower growth and better survival of leukaemic cells (Tzelepis *et al.*, 2016). A number of human malignancies are caused by mutation in more than one gene (4–5) and using conventional methods to treat such conditions is nearly impossible, however, using CRISPR/Cas9 and multiple gRNAs delivered by lentiviruses researchers modified five genes in a single haematopoietic cell (Heckl *et al.*, 2014).

Sickle cell disease (SCD) results from genetic mutations in the β-globin gene and are among the most common monogenic diseases in the world (Bauer and Orkin, 2015). Sickle cell anaemia is a genetic disease caused by homozygous mutation (A to T) in the 6th codon of the *HBB* gene which transcribes glutamic acid to valine, resulting in the production of abnormal β-globin and abnormal red blood cells production (Sun and Zhao, 2014). CRISPR/Cas9 corrected one *HBB* allele (proprotein convertase subtilisin) in high-quality iPSCs efficiently (Huang *et al.*, 2015).

The loss of function mutation in proprotein convertase subtilisin/kexin type 9 (*PCSK9*) is associated with low blood cholesterol level. Upon expression in liver this protein PCSK9 binds to the receptor and functions as an LDL receptor antagonist also degrading these receptors. This results in the lower uptake of LDL cholesterol. Using CRISPR/Cas9 the scientist disrupted *PCSK9* in mouse liver. Interestingly, the non-homologous end joining pathway (NHEJ) mediated gene disruption achieved 50% efficiency (Ding *et al.*, 2014).

Fanconi anaemia (FA) is a bone marrow failure syndrome caused by in-frame deletion in exon 4 of the *FANCC* gene, which produces congenital abnormalities and variation in chromosomal organization resulting in haematological and solid malignancies (Schifferli and Kuhne, 2015). The use of Cas9n corrected the mutation c.456+4A>T in the defective gene with higher efficiency (Osborn *et al.*, 2015).

It has been reported that 95% of polycythaemia vera (PV) is caused by a point mutation in *JAK2* (JAK2-V61F) (Levine and Gilliland, 2008) and a point mutation (AAT Z-mutation) the cause of α_1-antitrypsin (AAT deficiency) (Carrell and

Lomas, 2002). To compare the gene disruption ability of CRISPR/Cas9 and TALENs in myeloproliferative neoplasm polycythaemia vera (PV) derived iPSC to test the editing of target gene by HDR and gene disruption by NHEJ (Smith *et al.*, 2015). Cas9 induced a high frequency of indels produced as a result of DSB repair by NHE compared to that of TALENs. While the efficiency of genome editing based on HDR was comparable between CRISPR/Cas9 and TALENs.

CRISPR/Cas9 to treat movement disorders

Genetic disorders of movement or neurodegenerative disorders are caused by multiple genes and characterized by age-related accumulation of abnormal proteins causing degeneration of neurons through an unknown mechanism. Effective treatments of these diseases are of great importance. The most common neurodegenerative diseases are Alzheimer's disease (AD), Parkinson's disease (PD), amyotrophic lateral sclerosis (ALS), Huntington's disease (HD), and frontotemporal dementia (FTD). The genes coding for α-synuclein (*SNCA*) and parkin (*PRKN*) are the most important genes playing role in PD. Mostly, people with these mutations are prone to develop PD. The *SNCA* gene is crucial in PD as duplication of this locus accounts for 2% of familial cases (Lesage and Brice, 2009), because the α -synuclein protein is the main component of Lewy bodies (cytoplasmic misfolded proteins). Similarly in HD patients, on the other hand, there are aggregates or inclusions formed in an age-dependent manner by mutant huntingtin with an expanded polyQ tract (Li and Li, 2011).

There are several approaches for the treatment of these disorders based on gene silencing or gene delivery methods. These methods are in the pipeline of preclinical trials. In one study, the use of viral vector mediated Cas9 and gRNA delivery into mice brain cells using a fluorescent reporter gene to measure the effect on brain cells edited the targeted genes in neurons and astrocytes. When applied to the HD affected mice, the system efficiently reduced huntingtin (HT) accumulation (Talan, 2015). The technique led to 50% gene disruption in the test tube experiments. In another study CRISPR/Cas9 allowed knock-in designer receptors exclusively activated by designer drugs (DREADDs). The administration of clozapine-N-oxide (CNO) enables the precise regulation of human pluripotent stem cell (hPSC)-derived neurons. The transplantation of hPSC-derived human midbrain dopaminergic neurons into a Parkinson's disease mouse model rescues their motor function, and is able to be reversed or enhanced by using CNO (Chen *et al.*, 2016). These transplanted cells also caused behavioural changes in mice.

One of the most important applications of CRISPR/Cas9 is its application towards Duchenne muscular dystrophy (DMD). The disease is caused by mutation in the dystrophin gene at locus Xp21 on the short arm of the X chromosome (Ross *et al.*, 2005). Dystrophin is a protein present on the outermost layer of myofilament forming a connection between cytoskeleton and extracellular matrix. However, in its absence calcium enters into mitochondria and leads to capturing of water in the mitochondria, ultimately bursting.

Recently, CRISPR/Cas9 has been applied to shorten the dystrophin gene by removing the faulty exon in mice (Long *et al.*, 2016). The exon skipping technology with CRISPR/Cas9 can be applied to treat several other genetic diseases such as ataxia telangiectasia, congenital disorders of glycosylation and Niemann–Pick disease type C caused by errors in splicing.

Treating cystic fibrosis

Cystic fibrosis is characterized by accumulation of fluid in the gastrointestinal and pulmonary tract, leading to difficulties in breathing and recurrent infections. The gene responsible for the disease is *CFTR*, encoding a transmembrane conductance regulator. The function of this protein is to control the efflux and influx of mucus (epithelial fluid). However, the loss of function of this gene is due to deletion of three nucleotides, resulting in the loss of phenylalanine at 508th position of protein in about 70% of patients. CRISPR/Cas9-mediated gene editing allowed successful treatment of *CFTR* locus in intestinal stem cells from CF patients when transfected with a viral vector containing CRISPR components (Schwank *et al.*, 2013).

CRISPR/Cas9 to treat eye and ear diseases

Dystrophies in eyes are degenerative disorders of the eyes which, resulting from genetic heterogeneity, are characterized by symptoms such as night or colour blindness, abnormal vision and subsequent progression to complete and irreversible blindness. There are more than 120 genes associated with the development and function of eyes. Mutations in these genes are associated with different eye diseases. CRISPR may prove very fruitful to remediate these conditions because the eye is a constrained organ with relatively easy accessibility. To this end CRISPRs are also used to correct several genetic disorders of eyes, for instance retinitis pigmentosa. Retinitis pigmentosa is a serious disorder of the eye affecting approximately 1 in 4000 people in the USA. The disease is characterized by the breakdown of photoreceptor cells, resulting in gradual loss of vision, and has been recently edited successfully in iPSC for *RPGR* gene (Bassuk *et al.*, 2016). Similar application of CRISPR/Cas9 to treat the mutant gene *Crygc* has resulted in successful treatment of cataracts (Wu *et al.*, 2013).

Recently, a Japanese group has successfully applied CRISPR/Cas9 to treat transforming growth factor beta-induced (*TGFBI*) related corneal dystrophy using HDR pathway. The cultured corneal keratinocytes obtained from R124H granular dystrophy were transfected with gRNA and ssDNA HDR template *in vitro* (Usui, 2016).

The hearing loss diseases are mostly caused by mutation in a single gene, whereas few of them are caused by mutation in multiple genes (Rabionet *et al.*, 2000).

These genes are involved in different functions such as transcription factors, extracellular matrix molecules, cytoskeletal components, ion channels and transporters (cochlear molecules and hereditary deafness). The mutations in these genes are deletions, insertions, point mutations resulting in missense, nonsense or mutation of the responsible genes causing hearing disorder (http://hereditaryhearingloss.org/).

In order to develop a therapeutic strategy for a genetic disorder, understanding of the molecular mechanism and pathogenesis of the disease play a crucial role. The CRISPR/Cas9 technology platform generating a model organism or embryonic stem cells by mean of HDR or NHEJ that resembles the mutant condition can greatly help to achieve this goal.

A study has shown that a Cas9–sgRNA complex delivered by means of cationic lipid into the mouse inner hair cells *in vivo* showed efficient genome editing by knock-down of GFP signal in the Atoh1-GFP transgenic mice (Rabionet *et al.*, 2000). However, the development of a method to reduce the rapid degradation of Cas9 protein in cells and also the delivery of this complex to target the inner cells, for instance using supercharged protein to deliver Cas9-gRNA into inner cells also improving the frequency of HDR mediated gene editing to treat recessive mutation, will greatly facilitate new therapies for the treatment of genetic based deafness (Zou *et al.*, 2015).

Editing epigenome to treat cancer

The term 'epi (above)-genetics', means 'changes in the genome brought on by factors other than those related to conventional genetics', and represents a set of modifications that regulate gene expression during cell development. Epigenome editing, mainly comprising DNA methylation and histone modification, plays a crucial role in the regulation of gene expression (Jaenisch and Bird, 2003). The histone proteins which are involved in the packaging of DNA in the nucleus are modified by a number of epigenome modification including ubiquitination, phosphorylation, SUMOylation, acetylation. These reversible modification are carried out by specific enzymes (Arnaudo and Garcia, 2013). Chromatin folding and remodelling is also influenced by DNA methylation carried out by DNA methyltransferases (Jones and Takai, 2001; Reik and Walter, 2001).

Cancer is caused by changes in the epigenome, for instance in a recently reported study, the introduction of a genetic segment into the mouse genome upstream of gene p16 caused lung cancer, leukaemia or sarcomas in 27% mice compared to wild-type mice that did not induce cancer (Yu *et al.*, 2014). The gene p16 regulates cell division, and the exogenous segment used in this study was motifs from the human genome acting as a gene silencer during human development.

Methylation of cytosine plays a critical role in cancer pathogenesis. In cancer the methylation at

C is increased or decreased which in turn affects the regulatory sequences in DNA such as satellite DNA, repetitive sequences, and CpG sites. For example, mice with mutant DNA methyltransferases have low methylation in genome and developed lymphomas in T cells due to the activation of retroviral elements (Gaudet *et al.*, 2003; Howard *et al.*, 2008). However, in some human tumour conditions it is the overactivity of methyltransferases causing rapid proliferation of malignant cells (Issa, 2004).

Histone proteins are small proteins that work as a template to wrap the DNA to form the nucleosome. During the formation of the nucleosome the amino acids in the tail region (N-terminal) are exposed to enzyme activities responsible for epigenetic modification such as acetylation and phosphorylation, ubiquitination and methylation, and these modifications attribute to the tight or relaxed conformation of chromatin (Jenuwein and Allis, 2001; Strahl and Allis, 2000). The acetylation and phosphorylation are responsible for open chromatin structure as they represent the activation mark. The methylation may acts as either an activating mark (Strahl and Allis, 2000) or repressive mark (Czermin *et al.*, 2002; Müller *et al.*, 2002).

The exploration of the epigenome will provide a comprehensive understanding of its structure, function and association with diseases. To do so, dead Cas9 (dCas9) fused with a repressor, activator or modifier has been used. The fusion of dead Cas9 (double mutant Cas9 with no cleavage but binding ability) with Krüppel-associated box (KRAB) achieved suppression of targeted H2S enhancer region. The degree of genome-wide specificity and heterochromatin formation was not explored in this study. H2S enhancer, which enhances the expression of multiple globin genes using dCas9-KRAB trimethylation (H3K9me3) as the enhancer, was reported (Thakore *et al.*, 2015). The *Tyr* gene encoding for tyrosinase is a principal enzyme involved in the melanin biosynthesis pathway (Lavado and Montoliu, 2006; Olivares and Solano, 2009). Mutations in *Tyr* results in low melanin synthesis resulting in a genetic condition known as albinism. Using CRISPR/Cas9 guided by two gRNAs approach, a group of researchers deleted a 5′ region present ~12 kb 5′ region upstream of the mouse *Tyr* locus (Seruggia *et al.*, 2015). The resultant mice lacking the DNA sequence showed a clear phenotype. Histone acetyltransferase (HAT) catalytic core of the human acetyltransferase p300 is involved in a number of cellular processes (Chen and Li, 2011; Ogryzko *et al.*, 1996). Using dCas9 fused with catalytic histone acetyltransferase, acetylation of histone H3 lysine 27 at its target site was reported which significantly resulted in the regulation of epigenome and downstream gene expression (Hilton *et al.*, 2015).

The non-coding RNA (ncRNA) species, for instance micro-RNA (miRNA) AND short interfering RNA (siRNA), has been shown to regulate the epigenetics, thus regulating vital biological processes such as growth and development (Falahi *et al.*, 2015). In the prostate more than 100 regions are associated with the disease progression and recently more than 45 genes coding for non-coding RNAs are involved in the disease (Guo *et al.*, 2016). The targeting of these non-coding RNAs, especially miRNAs involved in cancer progression and development, could prove fruitful for cancer treatment.

CRISPR/Cas9 to generate disease models

The application of mouse models in regenerative medicine has played a tremendous role. The ability of Cas9 to edit genes by bringing point mutations or translocation at any locus in diverse cell lines has allowed the study of the function of many of these genes. Moreover, embryonic genome editing has paved the way to generate animal models to study diseases (Fig. 3.3). The benefit of the generation of these models is their resemblance with humans, thus allowing better understanding of the pathogenesis and progression of diseases, validating drug target sites and effective treatments, e.g. generation of a monkey model deficient in dystrophin gene has been shown to possess the same symptoms as a patient (Chen *et al.*, 2015). Similarly the system was recently applied to develop a pig model of Parkinson's disease by editing the responsible genes (Wang *et al.*, 2016a; Zhou *et al.*, 2015).

Huntington's disease (HD) is an age-dependent neuronal degenerative disorder caused by the abnormal repetition of trinucleotides (coding for glutamine residues) in the N-terminal of huntingtin (*htt*) gene (Gusella *et al.*, 1993). However, using nuclear transfer technology scientists generated a HD pig model expressing N-terminal mutant *htt* gene (Yang *et al.*, 2010). Using Cas9-gRNAs

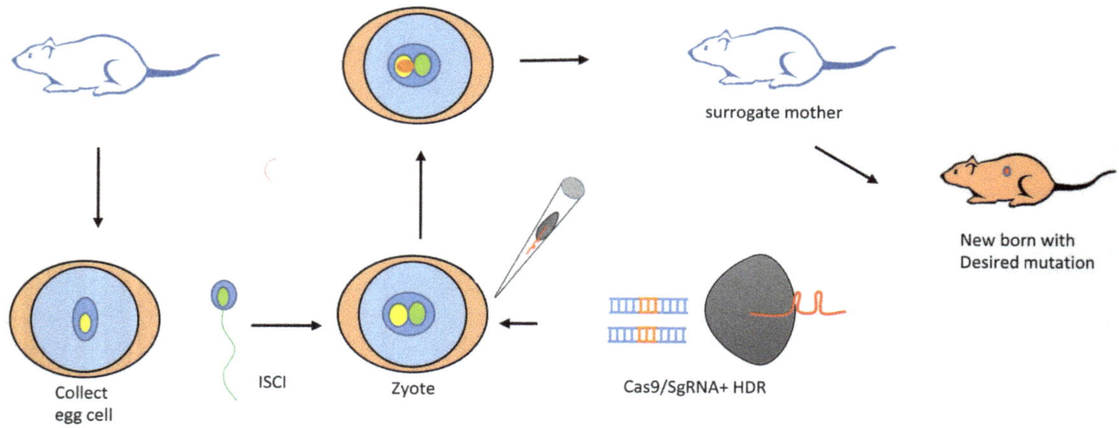

Figure 3.3 Representation of generating a transgenic mouse model using CRISPR mediated genome editing. The egg is collected from the donor, fertilized *in vitro*, and the zygote is engineered with CRISPR/Cas9 and implanted into surrogate mother. Post gestation mutant mice with desired mutations are born.

provided with exogenous desired mutant donor DNA fragments may be used to generate a model organism expressing the mutant gene.

The main hurdle to generating transgenic animal models is genetic mosaicism, arising due to the low rate and efficiency of Cas9 in inducing mutations. This may reflect the repression in translation of Cas9 in the zygote after microinjection of Cas9 mRNA and gRNA, but more likely the translation of Cas9 mRNA transcript is delayed after the first cell division (Oh *et al.*, 2000). In consistent to the previous discussion, the results of NHEJ repaired pathway introduce indels resulting in genetic mosaicism in transgenic rice. To this end efforts are made to supply single-cell fertilized embryos with purified Cas9 and sgRNA (Gratz *et al.*, 2013).

Genome editing of viruses

Hepatitis B virus (HBV) is a partially double-stranded DNA virus belonging to the Hepadnaviridae family. The chronic infection of HBV causes liver cirrhosis and carcinoma. Upon entry into the cell, the partially double-stranded DNA is converted into fully double-stranded DNA, i.e. covalently closed circular DNA (cccDNA) by viral polymerase. This cccDNA is highly stable in the cell and acts as a template for viral replication, thus making the removal of the virus challenging. Using Cas9 the researchers edited the viral genome by hydronamic injection of Cas9 and gRNA into the mouse tail that was already transfected with the HBV expression plasmid (Lin *et al.*, 2014a). Interestingly, the authors observed the cleavage of the viral vector and reduced surface antigen in the serum.

In a study carried out by Ebina *et al.* (2013), CRISPR/Cas9 was used to edit the HIV-integrated proviral DNA. Using gRNA to target HIV-1 long terminal repeat sequence at the 5′ and 3′ termini resulted in the subsequent suppression of the gene driven by LTR that in turn resulted in the inactivation of HIV replication in latently infected cells. This study also demonstrated the removal of viral DNA from the host chromosome. The use of dual gRNAs was shown effective to remove viral DNA fragment of hepatitis B (Wang *et al.*, 2015). These results indicate that the CRISPR/Cas9 technology can serve as a potential tool for clinical applications to cure infectious diseases. The use of this tool has been found to be effective in the clearance of latent viral DNA of hepatitis B virus, HIV and herpes virus (Hu *et al.*, 2014; Lin *et al.*, 2014a; Wang and Quake, 2014).

In a recent application of Cas9 to mutate HIV-1 integrated into cellular DNA, it was found that indel mutations resulted as a result of repair of Cas9-mediated DSB and impeded viral replication; however, some viruses escaped from CRISPR immunity (Wang *et al.*, 2016b). These results show that, although some of these indel mutations disrupt the recognition fidelity of Cas9, they are not lethal to the virus. Multiple targeting of viral DNA with CRISPR/Cas9 might help in effective antiviral therapy. The application of CRISPR/Cas9 to inactivate the viral genome has been carried out with a number of viruses including hepatitis B (Dong *et al.*, 2015; Karimova *et al.*, 2015; Lin *et al.*, 2014a; Ramanan *et al.*, 2015; Zhen *et al.*, 2015), Epstein–Barr (Wang and Quake, 2014; Yuen *et al.*, 2015), vaccina (Yuan *et al.*, 2015), adenovirus (Bi *et al.*, 2014), herpes simplex type 1 (Bi *et al.*, 2014) and human papillomavirus (Kennedy *et al.*, 2015). These applications proved that the CRISPR/Cas9 technology holds the capacity to be used as a potential tool for clinical applications to cure viral-based infectious diseases.

How to address the challenges associated with the application of CRISPR/Cas9 technology?

Programmable nuclease are guided by a stretch of guide sequence to target gene of interest, however, sometimes due to homology with secondary genomic sites, these nucleases also cause mutation in other sites known as off-target mutations. These mutations in some cases might have deleterious effects, for instance when applied in human clinical studies. The reason for such poor specificity is the fact that Cas9 uses a short stretch of 20 bp as a guide sequence in which 8–12 bp are highly crucial for targeting (Cong *et al.*, 2013). More possibly the requirement of 5′-NGG-3′5-′NAG-3′ as a PAM (protospacer adjacent motif) sequence also affects the recognition fidelity of Cas9. In order to overcome these problems there are various strategies developed by scientists to ameliorate the on-target specificity of Cas9, reviewed in recent comprehensive reviews (Jamal *et al.*, 2015; Tsai and Joung, 2016).

The development of safe and efficient delivery methods of Cas9–sgRNA complex to allow reduced toxicities and side effects is of paramount importance. Moreover, attempts should be made to avoid the random integration of donor DNA. The methods developed in last two decades can be practically applied for the delivery of Cas9-gRNA into cells (Kay, 2011; Mingozzi and High, 2011). Recently used adeno-associated virus shuttle vector is the optimum choice for the cargo method because of low cytotoxicity and immunogenicity (Ran *et al.*, 2015).

The fate of genome editing *in vivo* is different from *ex vivo*. The benefit of genome editing *ex vivo* is high chances to select cells with the corrected genotype without off-target mutation, expansion and its reintroduction into the patient's body. But the demerit of this approach is that during clonal expansion the cell may acquire unwanted genome alterations, and especially the induced pluripotent cells are more vulnerable to accumulate mutation and variation during expansion (Gore *et al.*, 2011; Ji *et al.*, 2012). In order to gain the genetic stability and avoid these unwanted mutations the culturing of stem cells in three-dimensional organoid cultures is beneficial (Huch *et al.*, 2015).

The low frequency of HDR-mediated genome editing *in vivo* compared with NHEJ is another hurdle associated with the therapeutic application of CRISPR/Cas9. During the cell cycle, NHEJ appears to be dominant during the S and G2 phases (Heyer *et al.*, 2010); HDR pathways, in contrast, are more prominent in the G1 and M phases (BP1, BRCA1 and the choice between recombination and end joining at DNA double-strand breaks). Cell cycle synchronization (Lin *et al.*, 2014b) and the use of an inhibitor (Chu *et al.*, 2015; Maruyama *et al.*, 2015) have been found to achieve a high frequency of HDR-mediated genome editing. The use of Cas9 nickase (Cas9n) with paired gRNAs has been actively used to promote HDR frequency (Ran *et al.*, 2013).

Conclusion and future directions

CRISPR/Cas9 is entering the era when it can revert the deleterious mutations causing genetic diseases at the DNA level. The scope is not limited to monogenic defects, where the strategy of using CRISPR/Cas9 is very straightforward, but extends to polygenic diseases, which can now be handled even at single-cell level. The power of CRISPR/Cas9 is no doubt very great, but its target-oriented use requires highly sophisticated clinical assays to e established to avoid any unwanted side-effects. The future of CRISPR/Cas9 genome editing is promising, and several laboratories are presently exploring its applications in wide areas of genome biology and personalized medicine.

Acknowledgements

This work was supported by the National Natural Science Foundation of China (Grant No.31371106,

31421064), Hubei Provincial Natural Science Foundation of China (Grant No. 36113011), Doctoral Fund of Ministry of Education of China (Grant No. 131012) and the Huazhong Agricultural University Scientific and Technological Self innovation Foundation (Program No. 52204-13002).

Conflict of interest

The authors disclose no conflict of interest.

References

Arnaudo, A.M., and Garcia, B.A. (2013). Proteomic characterization of novel histone post-translational modifications. Epigenetics & chromatin *6*, 1.

Barnes, D.E. (2001). Non-homologous end joining as a mechanism of DNA repair. Curr. Biol. *11*, R455–7.

Barrangou, R., Birmingham, A., Wiemann, S., Beijersbergen, R.L., Hornung, V., and van Brabant Smith, A. (2015). Advances in CRISPR-Cas9 genome engineering: lessons learned from RNA interference. Nucleic Acids Res. *43*, 3407–3419.

Bauer, D.E., and Orkin, S.H. (2015). Hemoglobin switching's surprise: the versatile transcription factor BCL11A is a master repressor of fetal hemoglobin. Curr. Opin. Genet. Dev.*33*, 62–70.

Bi, Y., Sun, L., Gao, D., Ding, C., Li, Z., Li, Y., Cun, W., and Li, Q. (2014). High-efficiency targeted editing of large viral genomes by RNA-guided nucleases. PLoS Pathog. *10*, e1004090.

Carrell, R.W., and Lomas, D.A. (2002). Alpha1-antitrypsin deficiency – a model for conformational diseases. N. Engl. J. Med. *346*, 45–53. http://dx.doi.org/10.1056/NEJMra010772

Chen, J., and Li, Q. (2011). Life and death of transcriptional co-activator p300. Epigenetics *6*, 957–961. http://dx.doi.org/10.4161/epi.6.8.16065

Chen, Y., Xiong, M., Dong, Y., Haberman, A., Cao, J., Liu, H., Zhou, W., and Zhang, S.-.C. (2016). Chemical control of Grafted human PSC-derived neurons in a mouse model of Parkinson's disease. Cell. Stem. Cell. *18*, 817–826. http://dx.doi.org/10.1016/j.stem.2016.03.014

Chen, Y., Zheng, Y., Kang, Y., Yang, W., Niu, Y., Guo, X., Tu, Z., Si, C., Wang, H., and Xing, R. (2015). Functional disruption of the dystrophin gene in rhesus monkey using CRISPR/Cas9. Hum. Mol. Genet. *24*, 3764–3774. http://dx.doi.org/10.1093/hmg/ddv120

Cho, S.W., Kim, S., Kim, J.M., and Kim, J.-.S. (2013). Targeted genome engineering in human cells with the Cas9 RNA-guided endonuclease. Nat. Biotechnol. *31*, 230–232. http://dx.doi.org/10.1038/nbt.2507

Chu, V.T., Weber, T., Wefers, B., Wurst, W., Sander, S., Rajewsky, K., and Kühn, R. (2015). Increasing the efficiency of homology-directed repair for CRISPR-Cas9-induced precise gene editing in mammalian cells. Nat. Biotechnol. *33*, 543–548. http://dx.doi.org/10.1038/nbt.3198

Cong, L., Ran, F.A., Cox, D., Lin, S., Barretto, R., Habib, N., Hsu, P.D., Wu, X., Jiang, W., and Marraffini, L.A. (2013). Multiplex genome engineering using CRISPR/Cas systems. Science *339*, 819–823.

Czermin, B., Melfi, R., McCabe, D., Seitz, V., Imhof, A., and Pirrotta, V. (2002). Drosophila enhancer of Zeste/ESC complexes have a histone H3 methyltransferase activity that marks chromosomal Polycomb sites. Cell *111*, 185–196.

Deveau, H., Barrangou, R., Garneau, J.E., Labonté, J., Fremaux, C., Boyaval, P., Romero, D.A., Horvath, P., and Moineau, S. (2008). Phage response to CRISPR-encoded resistance in *Streptococcus thermophilus*. J. Bacteriol. *190*, 1390–1400.

Ding, Q., Regan, S.N., Xia, Y., Oostrom, L.A., Cowan, C.A., and Musunuru, K. (2013). Enhanced efficiency of human pluripotent stem cell genome editing through replacing TALENs with CRISPRs. Cell. Stem. Cell. *12*, 393–394. http://dx.doi.org/10.1016/j.stem.2013.03.006

Ding, Q., Strong, A., Patel, K.M., Ng, S.-L., Gosis, B.S., Regan, S.N., Cowan, C.A., Rader, D.J., and Musunuru, K. (2014). Permanent alteration of PCSK9 with in vivo CRISPR-Cas9 genome editing. Circ. Res. *115*, 488–492. http://dx.doi.org/10.1161/CIRCRESAHA.115.304351

Dong, C., Qu, L., Wang, H., Wei, L., Dong, Y., and Xiong, S. (2015). Targeting hepatitis B virus cccDNA by CRISPR/Cas9 nuclease efficiently inhibits viral replication. Antiviral. Res. *118*, 110–117. http://dx.doi.org/10.1016/j.antiviral.2015.03.015

Doudna, J.A., and Charpentier, E. (2014). The new frontier of genome engineering with CRISPR-Cas9. Science *346*, 1258096. http://dx.doi.org/10.1126/science.1258096

Dudás, A., and Chovanec, M. (2004). DNA double-strand break repair by homologous recombination. Mutat. Res. *566*, 131–167. http://dx.doi.org/10.1016/j.mrrev.2003.07.001

Ebina, H., Misawa, N., Kanemura, Y., and Koyanagi, Y. (2013). Harnessing the CRISPR/Cas9 system to disrupt latent HIV-1 provirus. Scientific Reports 3.

Falahi, F., Sgro, A., and Blancafort, P. (2015). Epigenome engineering in cancer: fairytale or a realistic path to the clinic? Front. Oncol. *5*, 22. http://dx.doi.org/10.3389/fonc.2015.00022

Gaudet, F., Hodgson, J.G., Eden, A., Jackson-Grusby, L., Dausman, J., Gray, J.W., Leonhardt, H., and Jaenisch, R. (2003). Induction of tumors in mice by genomic hypomethylation. Science *300*, 489–492. http://dx.doi.org/10.1126/science.1083558

Gore, A., Li, Z., Fung, H.-L., Young, J.E., Agarwal, S., Antosiewicz-Bourget, J., Canto, I., Giorgetti, A., Israel, M.A., and Kiskinis, E. (2011). Somatic coding mutations in human induced pluripotent stem cells. Nature *471*, 63–67. http://dx.doi.org/10.1038/nature09805

Gratz, S.J., Cummings, A.M., Nguyen, J.N., Hamm, D.C., Donohue, L.K., Harrison, M.M., Wildonger, J., and O'Connor-Giles, K.M. (2013). Genome engineering of *Drosophila* with the CRISPR RNA-guided Cas9 nuclease. Genetics *194*, 1029–1035. http://dx.doi.org/10.1534/genetics.113.152710

Guo, H., Ahmed, M., Zhang, F., Yao, C.Q., Li, S., Liang, Y., Hua, J., Soares, F., Sun, Y., and Langstein, J. (2016). Modulation of long noncoding RNAs by risk SNPs underlying genetic predispositions to prostate cancer. Nat. Gen. *48*, 1142–1500.

Gusella, J.F., MacDonald, M.E., Ambrose, C.M., and Duyao, M.P. (1993). Molecular genetics of Huntington's disease. Arch. Neurol. *50*, 1157–1163.

Heckl, D., Kowalczyk, M.S., Yudovich, D., Belizaire, R., Puram, R.V., McConkey, M.E., Thielke, A., Aster, J.C., Regev, A., and Ebert, B.L. (2014). Generation of mouse models of myeloid malignancy with combinatorial genetic lesions using CRISPR-Cas9 genome editing. Nat. Biotechnol. *32*, 941–946.

Heyer, W.-D., Ehmsen, K.T., and Liu, J. (2010). Regulation of homologous recombination in eukaryotes. Annu. Rev. Gen. *44*, 113.

Hilton, I.B., D'Ippolito, A.M., Vockley, C.M., Thakore, P.I., Crawford, G.E., Reddy, T.E., and Gersbach, C.A. (2015). Epigenome editing by a CRISPR-Cas9-based acetyltransferase activates genes from promoters and enhancers. Nat. Biotechnol. *33*, 510–517. http://dx.doi.org/10.1038/nbt.3199

Howard, G., Eiges, R., Gaudet, F., Jaenisch, R., and Eden, A. (2008). Activation and transposition of endogenous retroviral elements in hypomethylation induced tumors in mice. Oncogene *27*, 404–408.

Hu, W., Kaminski, R., Yang, F., Zhang, Y., Cosentino, L., Li, F., Luo, B., Alvarez-Carbonell, D., Garcia-Mesa, Y., and Karn, J.,(2014). RNA-directed gene editing specifically eradicates latent and prevents new HIV-1 infection. Proc. Natl. Acad. Sci. U.S.A. *111*, 11461–11466. http://dx.doi.org/10.1073/pnas.1405186111

Huang, X., Wang, Y., Yan, W., Smith, C., Ye, Z., Wang, J., Gao, Y., Mendelsohn, L., and Cheng, L. (2015). Production of Gene corrected adult beta globin protein in human erythrocytes differentiated from patient iPSCs After genome editing of the sickle point mutation. Stem Cells *33*, 1470–1479.

Huch, M., Gehart, H., van Boxtel, R., Hamer, K., Blokzijl, F., Verstegen, M.M., Ellis, E., van Wenum, M., Fuchs, S.A., and de Ligt, J. (2015). Long-term culture of genome-stable bipotent stem cells from adult human liver. Cell *160*, 299–312. http://dx.doi.org/10.1016/j.cell.2014.11.050

Ishino, Y., Shinagawa, H., Makino, K., Amemura, M., and Nakata, A. (1987). Nucleotide sequence of the iap gene, responsible for alkaline phosphatase isozyme conversion in *Escherichia coli*, and identification of the gene product. J. Bacteriol. *169*, 5429–5433.

Issa, J.-P. (2004). CpG island methylator phenotype in cancer. Nat. Rev. Cancer *4*, 988–993.

Jaenisch, R., and Bird, A. (2003). Epigenetic regulation of gene expression: how the genome integrates intrinsic and environmental signals. Nat. Genet. *33*, 245–254. http://dx.doi.org/10.1038/ng1089

Jamal, M., Khan, F.A., Da, L., Habib, Z., Dai, J., and Cao, G. (2015). Keeping CRISPR/Cas on-target. Curr Issues Mol Biol *20*, 1-20.

Jenuwein, T., and Allis, C.D. (2001). Translating the histone code. Science *293*, 1074–1080.

Ji, J., Ng, S.H., Sharma, V., Neculai, D., Hussein, S., Sam, M., Trinh, Q., Church, G.M., Mcpherson, J.D., and Nagy, A. (2012). Elevated coding mutation rate during the reprogramming of human somatic cells into induced pluripotent stem cells. Stem Cells *30*, 435–440. http://dx.doi.org/10.1002/stem.1011

Jinek, M., Chylinski, K., Fonfara, I., Hauer, M., Doudna, J.A., and Charpentier, E. (2012). A programmable dual-RNA-guided DNA endonuclease in adaptive bacterial immunity. Science *337*, 816–821. http://dx.doi.org/10.1126/science.1225829

Jones, P.A., and Takai, D. (2001). The role of DNA methylation in mammalian epigenetics. Science *293*, 1068–1070. http://dx.doi.org/10.1126/science.1063852

Joung, J.K., and Sander, J.D. (2013). TALENs: a widely applicable technology for targeted genome editing. Nat. Rev. Mol. Cell. Biol. *14*, 49–55. http://dx.doi.org/10.1038/nrm3486

Karimova, M., Beschorner, N., Dammermann, W., Chemnitz, J., Indenbirken, D., Bockmann, J.-H., Grundhoff, A., Lüth, S., Buchholz, F., and zur Wiesch, J.S. (2015). CRISPR/Cas9 nickase-mediated disruption of hepatitis B virus open reading frame S and X. Sci. Rep. *5*.

Kay, M.A. (2011). State-of-the-art gene-based therapies: the road ahead. Nat. Rev. Genet. *12*, 316–328. http://dx.doi.org/10.1038/nrg2971

Kennedy, E.M., Bassit, L.C., Mueller, H., Kornepati, A.V., Bogerd, H.P., Nie, T., Chatterjee, P., Javanbakht, H., Schinazi, R.F., and Cullen, B.R. (2015). Suppression of hepatitis B virus DNA accumulation in chronically infected cells using a bacterial CRISPR/Cas RNA-guided DNA endonuclease. Virology *476*, 196-205.

Koonin, E.V., and Makarova, K.S. (2009). CRISPR-Cas: an adaptive immunity system in prokaryotes. F1000. Biol. Rep. *1*, 95. http://dx.doi.org/10.3410/B1-95

Lavado, A., and Montoliu, L. (2006). New animal models to study the role of tyrosinase in normal retinal development. Front. Biosci. *11*, 743–752.

Lesage, S., and Brice, A. (2009). Parkinson's disease: from monogenic forms to genetic susceptibility factors. Human molecular genetics *18*, R48-R59.

Levine, R.L., and Gilliland, D.G. (2008). Myeloproliferative disorders. Blood *112*, 2190–2198. http://dx.doi.org/10.1182/blood-2008-03-077966

Li, X.-J., and Li, S. (2011). Proteasomal dysfunction in aging and Huntington disease. Neurobiol. Dis. *43*, 4–8. http://dx.doi.org/10.1016/j.nbd.2010.11.018

Liang, P., Xu, Y., Zhang, X., Ding, C., Huang, R., Zhang, Z., Lv, J., Xie, X., Chen, Y., and Li, Y. (2015). CRISPR/Cas9-mediated gene editing in human tripronuclear zygotes. Protein & Cell *6*, 363–372. http://dx.doi.org/10.1007/s13238-015-0153-5

Lin, S.-R., Yang, H.-C., Kuo, Y.-T., Liu, C.-J., Yang, T.-Y., Sung, K.-C., Lin, Y.-Y., Wang, H.-Y., Wang, C.-C., and Shen, Y.-C. (2014a). The CRISPR/Cas9 system facilitates clearance of the intrahepatic HBV templates in vivo. Mol. Ther. Nucleic. Acids. *3*, e186. http://dx.doi.org/10.1038/mtna.2014.38.

Lin, S., Staahl, B.T., Alla, R.K., and Doudna, J.A. (2014b). Enhanced homology-directed human genome engineering by controlled timing of CRISPR/Cas9 delivery. Elife *3*, e04766. http://dx.doi.org/10.7554/eLife.04766

Long, C., Amoasii, L., Mireault, A.A., McAnally, J.R., Li, H., Sanchez-Ortiz, E., Bhattacharyya, S., Shelton, J.M., Bassel-Duby, R., and Olson, E.N. (2016). Postnatal genome editing partially restores dystrophin expression in a mouse model of muscular dystrophy. Science *351*, 400–403.

Long, C., McAnally, J.R., Shelton, J.M., Mireault, A.A., Bassel-Duby, R., and Olson, E.N. (2014). Prevention of muscular dystrophy in mice by CRISPR/Cas9-mediated editing of germline DNA. Science *345*, 1184–1188. http://dx.doi.org/10.1126/science.1254445

Maruyama, T., Dougan, S.K., Truttmann, M.C., Bilate, A.M., Ingram, J.R., and Ploegh, H.L. (2015). Increasing the efficiency of precise genome editing with CRISPR-Cas9 by inhibition of nonhomologous end joining. Nat. Biotechnol. *33*, 538-542.

Mingozzi, F., and High, K.A. (2011). Therapeutic in vivo gene transfer for genetic disease using AAV: progress and challenges. Nat. Rev. Genet. *12*, 341–355. http://dx.doi.org/10.1038/nrg2988

Müller, J., Hart, C.M., Francis, N.J., Vargas, M.L., Sengupta, A., Wild, B., Miller, E.L., O'Connor, M.B., Kingston, R.E., and Simon, J.A. (2002). Histone methyltransferase activity of a Drosophila Polycomb group repressor complex. Cell *111*, 197–208.

Niu, Y., Shen, B., Cui, Y., Chen, Y., Wang, J., Wang, L., Kang, Y., Zhao, X., Si, W., and Li, W. (2014). Generation of gene-modified cynomolgus monkey via Cas9/RNA-mediated gene targeting in one-cell embryos. Cell *156*, 836–843.

Ogryzko, V.V., Schiltz, R.L., Russanova, V., Howard, B.H., and Nakatani, Y. (1996). The transcriptional coactivators p300 and CBP are histone acetyltransferases. Cell *87*, 953–959.

Oh, B., Hwang, S., McLaughlin, J., Solter, D., and Knowles, B.B. (2000). Timely translation during the mouse oocyte-to-embryo transition. Development *127*, 3795–3803.

Olivares, C., and Solano, F. (2009). New insights into the active site structure and catalytic mechanism of tyrosinase and its related proteins. Pigment. Cell. Melanoma Res. *22*, 750–760. http://dx.doi.org/10.1111/j.1755-148X.2009.00636.x

Osborn, M.J., Gabriel, R., Webber, B.R., DeFeo, A.P., McElroy, A.N., Jarjour, J., Starker, C.G., Wagner, J.E., Joung, J.K., and Voytas, D.F. (2015). Fanconi anemia gene editing by the CRISPR/Cas9 system. Hum. Gene Ther. *26*, 114–126. http://dx.doi.org/10.1089/hum.2014.111

Perry, A.C., Rothman, A., Jose, I., Feinstein, P., Mombaerts, P., Cooke, H.J., and Wakayama, T. (2001). Efficient metaphase II transgenesis with different transgene archetypes. Nat. Biotechnol. *19*, 1071–1073. http://dx.doi.org/10.1038/nbt1101-1071.

Rabionet, R., Gasparini, P., and Estivill, X. (2000). Molecular genetics of hearing impairment due to mutations in gap junction genes encoding beta connexins. Human Mutat. *16*, 190.

Ramanan, V., Shlomai, A., Cox, D.B., Schwartz, R.E., Michailidis, E., Bhatta, A., Scott, D.A., Zhang, F., Rice, C.M., and Bhatia, S.N. (2015). CRISPR/Cas9 cleavage of viral DNA efficiently suppresses hepatitis B virus. Sci. Rep. *5*, 10833. http://dx.doi.org/10.1038/srep10833

Ran, F.A., Cong, L., Yan, W.X., Scott, D.A., Gootenberg, J.S., Kriz, A.J., Zetsche, B., Shalem, O., Wu, X., and Makarova, K.S. (2015). In vivo genome editing using Staphylococcus aureus Cas9. Nature *520*, 186–191. http://dx.doi.org/10.1038/nature14299

Ran, F.A., Hsu, P.D., Lin, C.-Y., Gootenberg, J.S., Konermann, S., Trevino, A.E., Scott, D.A., Inoue, A., Matoba, S., and Zhang, Y. (2013). Double nicking by RNA-guided CRISPR Cas9 for enhanced genome editing specificity. Cell *154*, 1380–1389. http://dx.doi.org/10.1016/j.cell.2013.08.021.

Reik, W., and Walter, J. (2001). Genomic imprinting: parental influence on the genome. Nat. Rev. Genet. *2*, 21–32. http://dx.doi.org/10.1038/35047554

Ross, M.T., Grafham, D.V., Coffey, A.J., Scherer, S., McLay, K., Muzny, D., Platzer, M., Howell, G.R., Burrows, C., and Bird, C.P. (2005). The DNA sequence of the human X chromosome. Nature *434*, 325–337.

Schwank, G., Koo, B.-K., Sasselli, V., Dekkers, Johanna F., Heo, I., Demircan, T., Sasaki, N., Boymans, S., Cuppen, E., van der Ent, Cornelis, K., *et al.* (2013). Functional repair of CFTR by CRISPR/Cas9 in intestinal stem cell organoids of cystic fibrosis patients. Cell. Stem. Cell. *13*, 653–658. http://dx.doi.org/10.1016/j.stem.2013.11.002

Seruggia, D., Fernández, A., Cantero, M., Pelczar, P., and Montoliu, L. (2015). Functional validation of mouse tyrosinase non-coding regulatory DNA elements by CRISPR–Cas9-mediated mutagenesis. Nucleic Acids Res. *43*, 4855–4867. http://dx.doi.org/10.1093/nar/gkv375

Smith, C., Abalde-Atristain, L., He, C., Brodsky, B.R., Braunstein, E.M., Chaudhari, P., Jang, Y.-Y., Cheng, L., and Ye, Z. (2015). Efficient and allele-specific genome editing of disease loci in human iPSCs. Mol. Ther. *23*, 570–577. http://dx.doi.org/10.1038/mt.2014.226

Strahl, B.D., and Allis, C.D. (2000). The language of covalent histone modifications. Nature *403*, 41–45. http://dx.doi.org/10.1038/47412

Sun, N., and Zhao, H. (2014). Seamless correction of the sickle cell disease mutation of the HBB gene in human induced pluripotent stem cells using TALENs. Biotechnol. Bioeng. *111*, 1048–1053. http://dx.doi.org/10.1002/bit.25018

Suzuki, T., Asami, M., and Perry, A.C. (2014). Asymmetric parental genome engineering by Cas9 during mouse meiotic exit. Scientific Reports *4*.

Talan, J. (2015). News from the Society for Neuroscience Annual Meeting: Gene Editing Techniques Show Promise in Silencing or Inhibiting the Mutant Huntington's Disease Gene. Neurol. Today *15*, 14–16.

Thakore, P.I., D'Ippolito, A.M., Song, L., Safi, A., Shivakumar, N.K., Kabadi, A.M., Reddy, T.E., Crawford, G.E., and Gersbach, C.A. (2015). Highly specific epigenome editing by CRISPR-Cas9 repressors for silencing of distal regulatory elements. Nat. Methods. *12*, 1143–1149.

Tsai, S.Q., and Joung, J.K. (2016). Defining and improving the genome-wide specificities of CRISPR-Cas9 nucleases. Nat. Rev. Genet. *17*, 300–312. http://dx.doi.org/10.1038/nrg.2016.28

Tzelepis, K., Koike-Yusa, H., De Braekeleer, E., Li, Y., Metzakopian, E., Dovey, O.M., Mupo, A., Grinkevich, V., Li, M., and Mazan, M. (2016). A CRISPR dropout screen identifies genetic vulnerabilities and therapeutic targets in acute myeloid leukemia. Cell. Rep. *17*, 1193–1205.

Urnov, F.D., Rebar, E.J., Holmes, M.C., Zhang, H.S., and Gregory, P.D. (2010). Genome editing with engineered zinc finger nucleases. Nat. Rev. Genet. *11*, 636–646. http://dx.doi.org/10.1038/nrg2842

Usui, T. (2016). [To Protect Corneal Transparency against Diseases]. Nippon Ganka Gakkai Zasshi *120*, 246–262; discussion 263.

Wang, H., Yang, H., Shivalila, C.S., Dawlaty, M.M., Cheng, A.W., Zhang, F., and Jaenisch, R. (2013). One-step generation of mice carrying mutations in multiple genes by CRISPR/Cas-mediated genome engineering. Cell *153*, 910–918. http://dx.doi.org/10.1016/j.cell.2013.04.025

Wang, J., and Quake, S.R. (2014). RNA-guided endonuclease provides a therapeutic strategy to cure latent herpesviridae infection. Proc. Natl. Acad. Sci. U.S.A. *111*, 13157-13162.

Wang, J., Xu, Z.-W., Liu, S., Zhang, R.-Y., Ding, S.-L., Xie, X.-M., Long, L., Chen, X.-M., Zhuang, H., and Lu, F.-M. (2015). Dual gRNAs guided CRISPR/Cas9 system inhibits hepatitis B virus replication. World J. Gastroenterol. *21*, 9554.

Wang, X., Cao, C., Huang, J., Yao, J., Hai, T., Zheng, Q., Wang, X., Zhang, H., Qin, G., and Cheng, J. (2016a). One-step generation of triple gene-targeted pigs using CRISPR/Cas9 system. Scientific Rep. *6*.

Wang, Z., Pan, Q., Gendron, P., Zhu, W., Guo, F., Cen, S., Wainberg, M.A., and Liang, C. (2016b). CRISPR/Cas9-derived mutations both inhibit HIV-1 replication and accelerate viral escape. Cell Rep. *15*, 481–489.

Wiedenheft, B., Sternberg, S.H., and Doudna, J.A. (2012). RNA-guided genetic silencing systems in bacteria and archaea. Nature *482*, 331–338. http://dx.doi.org/10.1038/nature10886

Wu, Y., Liang, D., Wang, Y., Bai, M., Tang, W., Bao, S., Yan, Z., Li, D., and Li, J. (2013). Correction of a genetic disease in mouse via use of CRISPR-Cas9. Cell. Stem. Cell. *13*, 659–662. http://dx.doi.org/10.1016/j.stem.2013.10.016

Yanagimachi, R. (2002). Mammalian transgenesis by intracytoplasmic sperm injection (Google Patents).

Yang, D., Wang, C.-E., Zhao, B., Li, W., Ouyang, Z., Liu, Z., Yang, H., Fan, P., O'Neill, A., and Gu, W. (2010). Expression of Huntington's disease protein results in apoptotic neurons in the brains of cloned transgenic pigs. Hum. Mol. Genet. *19*, 3983–3994. http://dx.doi.org/10.1093/hmg/ddq313

Yu, D.-H., Waterland, R.A., Zhang, P., Schady, D., Chen, M.-H., Guan, Y., Gadkari, M., and Shen, L. (2014). Targeted p16Ink4a epimutation causes tumorigenesis and reduces survival in mice. J. Clin. Invest. *124*, 3708–3712. http://dx.doi.org/10.1172/JCI76507

Yuan, M., Zhang, W., Wang, J., Al Yaghchi, C., Ahmed, J., Chard, L., Lemoine, N.R., and Wang, Y. (2015). Efficiently editing the vaccinia virus genome by using the CRISPR-Cas9 system. J. Virol. *89*, 5176–5179. http://dx.doi.org/10.1128/JVI.00339-15

Yuen, K.-S., Chan, C.-P., Wong, N.-H.M., Ho, C.-H., Ho, T.-H., Lei, T., Deng, W., Tsao, S.W., Chen, H., and Kok, K.-H. (2015). CRISPR/Cas9-mediated genome editing of Epstein–Barr virus in human cells. J. Gen. Virol. *96*, 626–636. http://dx.doi.org/10.1099/jgv.0.000012

Zhen, S., Hua, L., Liu, Y., Gao, L., Fu, J., Wan, D., Dong, L., Song, H., and Gao, X. (2015). Harnessing the clustered regularly interspaced short palindromic repeat (CRISPR)/CRISPR-associated Cas9 system to disrupt the hepatitis B virus. Gene. Ther. *22*, 404–412. http://dx.doi.org/10.1038/gt.2015.2

Zhou, X., Xin, J., Fan, N., Zou, Q., Huang, J., Ouyang, Z., Zhao, Y., Zhao, B., Liu, Z., and Lai, S. (2015). Generation of CRISPR/Cas9-mediated gene-targeted pigs via somatic cell nuclear transfer. Cell. Mol. Life. Sci. *72*, 1175–1184. http://dx.doi.org/10.1007/s00018-014-1744-7

Zou, B., Mittal, R., Grati, M.h., Lu, Z., Shu, Y., Tao, Y., Feng, Y., Xie, D., Kong, W., and Yang, S. (2015). The application of genome editing in studying hearing loss. Hear. Res. *327*, 102–108. http://dx.doi.org/10.1016/j.heares.2015.04.016

An Era of CRISPR/Cas9-mediated Plant Genome Editing

4

Haris Khurshid[1,2], Sohail Ahmad Jan[1,3*], Zabta Khan Shinwari[1], Muhammad Jamal[4] and Sabir Hussain Shah[5]

[1]Department of Biotechnology, Quaid-i-Azam University, Islamabad, Pakistan.
[2]Oilseeds Research Program, National Agricultural Research Centre, Islamabad, Pakistan.
[3]Plant Genetic Resources Institute, National Agricultural Research Centre, Islamabad, Pakistan.
[4]State Key Laboratory of Agricultural Microbiology, College of Animal Science and Medicine, Huazhong Agricultural University, Wuhan, The People's Republic of China.
[5]Department of Agricultural Sciences, Allama Iqbal Open University, Islamabad, Pakistan.

*Correspondence: sjan.parc@gmail.com

https://doi.org/10.21775/9781910190630.04

Abstract

Recently the engineered nucleases have revolutionized genome editing to perturb gene expression at specific sites in complex eukaryotic genomes. Three important classes of these genome editing tools are Zinc Finger Nucleases (ZFN), Meganucleases and Transcription Activator-Like Effector Nucleases (TALEN) which work as hybrid systems comprising of target-specific DNA binding domains and molecular scissors or restriction endonucleases. Moreover, the more recent type II clustered regularly inter-spaced short palindromic repeats (CRISPR)-associated protein (CRISPR/Cas9) system has become the favourite plant genome editing tool for its precision and RNA based specificity unlike its counterparts which rely on protein based specificity. Plasmid-mediated co-delivery of multiple sgRNAs and Cas9 to the plant cell can simultaneously alter more than one target loci which enables multiplex genome editing. In this review, we discuss recent advancements in the CRISPR/Cas9 technology mechanism, theory and its applications in plants and agriculture. We also suggest that the CRISPR/Cas9 as an effective genome editing tool, has vast potential for crop improvement and studying gene regulation mechanisms and chromatin remodelling.

CRISPR/Cas9: ultimate tool for multiplex genome editing in plants

The small size of Cas9 and the relative ease of delivery of Cas9 and multiple sgRNAs to the cell makes it possible to simultaneously alter more than one target loci in a single genome, which is known as multiplexing or multiplex genome editing. On the contrary, the larger size of ZFN and TALEN which requires a pair of proteins recognizing complementary strands of DNA for introducing DSBs makes these techniques less suitable for multi locus gene editing (Bortesi and Fischer, 2015). Moreover, a number of studies exploited the use of two or more sgRNAs for multiplex genome editing in the model plant *Arabidopsis* (Li *et al.*, 2013; Mao *et al.*, 2013), rice (Zhang *et al.*, 2014) and tomato (Brooks *et al.*, 2014). Through multiplex genome editing chromosomal deletions from a few hundred base pairs up to tens of thousands have been made in *N. benthamiana* (Belhaj *et al.*, 2013), *Arabidopsis* (Mao *et al.*, 2013; Li *et al.*, 2013) and rice (Zhou *et al.*, 2014). This enables researchers to delete entire cluster of genes by deleting chromosomal fragments.

Recently methods have been proposed (Xing *et al.*, 2014) for developing gRNA modules and configuring expression cassettes having multiple

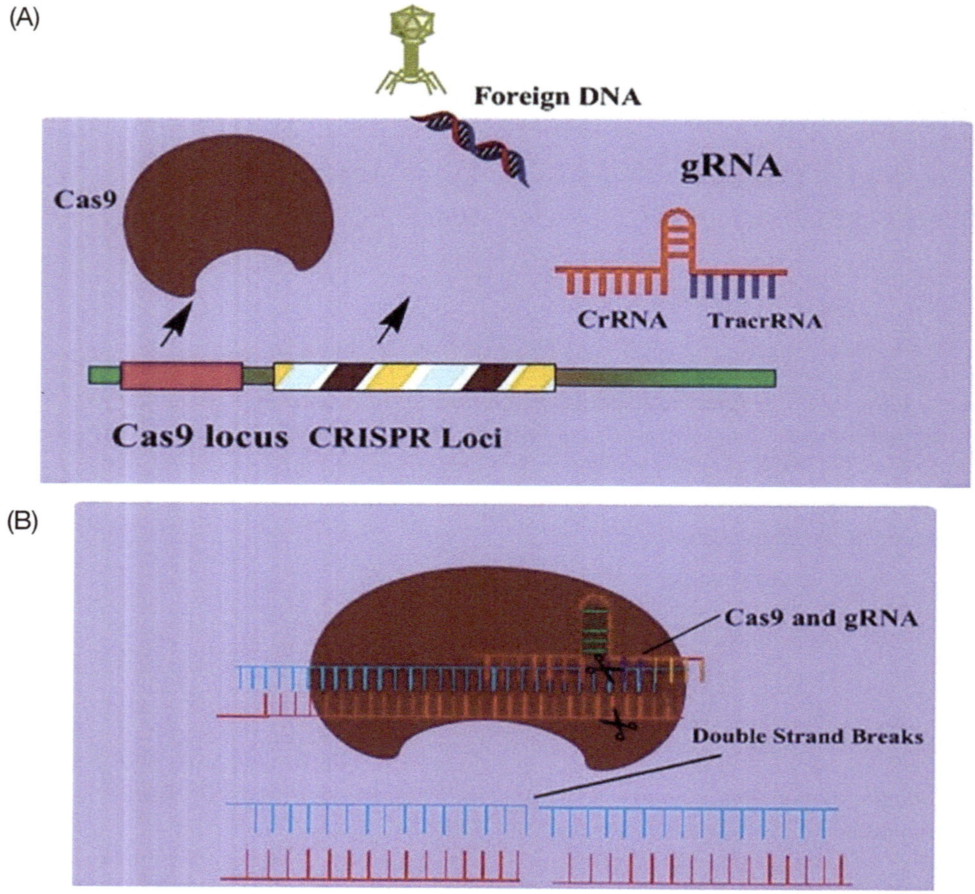

Figure 4.1 (A) Bacteria and archaea have genetic defence mechanisms against the invading DNA of bacteriophages. Their genome has Cas protein as well as CRISPR loci, which upon invasion by foreign DNA transcribe respective Cas proteins and guide RNA. Guide RNA or gRNA is a chimera formed by CRISPR RNA and trans-activating CRISPR RNA or tracrRNA. (B) Cas9/gRNA complex intercept target DNA which is complementary to gRNA and restriction sites are recognized through PAM sequence, i.e. 5′-NGG-3 located adjacent to the target sequence. Once localized, the RuvC and HNH sub-domains of NUC domain of Cas9 start endonuclease activity by inducing double stranded breaks (DSBs) in the target DNA.

gRNAs in plants. The application of these broad range resources comprised of binary vectors and gRNA are highly compatible with the requirements of a number of plants systems under complex conditions. This offers researchers with capability not only to customize their gRNA module according to specific plant systems but also to employ multiple gRNAs in a single cassette for multiplex genome editing.

Applications in plants

Since its discovery as modern genome editing tool, numerous research studies have evaluated CRISPR/Cas9 Type II system's possible role in gene expression regulation, gene knock-in/-out, site-specific mutation and epigenetic mechanisms in various organisms (Figs. 4.1 and 4.2). Moreover, its application in plants and agriculture are gradually becoming the focus of researchers as the tool offers new and efficient ways of manipulating a plant's genome to meet a broad set of enforced conditions.

As a model plant, vast genomic data of *Arabidopsis* have made this species an ideal target for CRISPR/Cas9 based genome editing and number of genes including *ADH1* (Schiml *et al.*, 2014), *AtPDS3*, *AtFLS2* (Li *et al.*, 2013), *ADH1* and *TT4*

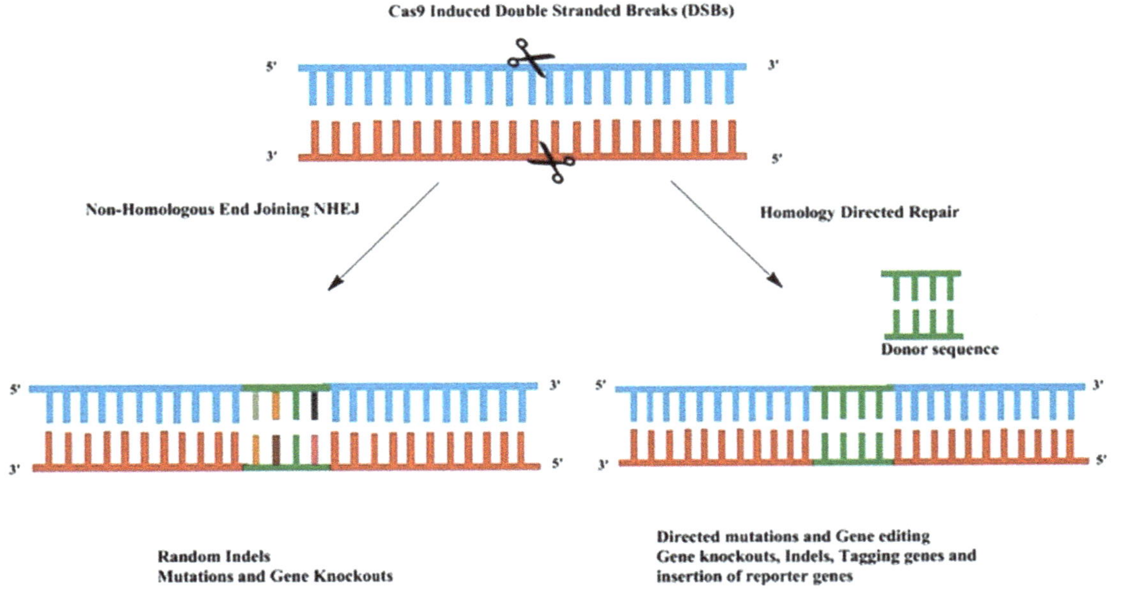

Figure 4.2 In a cell, after double-stranded breaks (DSBs) are introduced, the damaged DNA can be repaired by one of the two automated DNA repair pathways, i.e. the non-homologous end joining (NHEJ), which is more error prone often resulting in insertion deletions (indels), and can be exploited for frameshift mutation to cause gene knockouts. The second is homology directed repair (HDR) pathway which is comparatively precise and repairs DNA in nearly perfect way. In the presence of a user-supplied donor template HDR can be utilized for gene knock-in, gene correction or sequence replacement.

genes (Fauser *et al.*, 2014), *Flowering Locus T (FT)* and *Squamosa Promoter Binding Protein-Like 4* genes (Hyun *et al.*, 2015), and *BRI1, GAI, CHL1* (Mao *et al.*, 2013; Feng *et al.*, 2013) have been targeted. The resultant targeted mutations had varying degrees of efficiency from 1.1% to 85% in T_1 but were found to be stably heritable with as high a percentage as 79.4% (Feng *et al.*, 2013). CRISPR/Cas9-mediated targeted mutations have been induced using two sgRNA expression cassettes to perturb *Chlorophyll A Oxygenase1* and *LAZY1* (Mao *et al.*, 2013). This system has been used in combination with VIGS (virus-induced gene silencing) to study transient expression system in tobacco (*Nicotiana benthamiana*) (Ali *et al.*, 2015; Nekrasov *et al.*, 2013; Gao *et al.*, 2015). In tomato successful silencing of *Argonaute 7* has been achieved through this system resulting in obvious morphological changes in its leaves (Brooks *et al.*, 2014). A number of loci have been successfully targeted for gene perturbation and chromosomal deletion in rice (Jiang *et al.*, 2013; Shan *et al.*, 2014; Miao *et al.*, 2013; Zhou *et al.*, 2014). In rice, *OsPDS* and *OsBADH2* gene knockouts were obtained with mutation rates as high as 9.4% and 7.1%, respectively (Shan *et al.*, 2014). More recently, through improved Cas9/sgRNA platforms, biallelic and homozygous mutations were introduced with a maximum of 85.4% mutation rate (Ma *et al.*, 2015). Site-specific mutations have been effectively introduced through sgRNA and Cas9 delivery into rice protoplast to target promoter regions of *OsSWEET11* and *OsSWEET14*, bacterial blight susceptibility genes (Jiang *et al.*, 2013). Similarly, CRISPR/Cas9 mediated mutations were induced in maize protoplast gene *ZmIPK* (Liang *et al.*, 2014), wheat *LOX2* gene (Shan *et al.*, 2014) and sorghum (*DsRED2*) (Jiang *et al.*, 2013). Furthermore, sgRNA/Cas9 delivery protocol has been devised in important food crops with complex genome like hexaploid wheat, where gene knock-out in a single set of chromosomes may not be sufficient to produce the intended phenotype (Shan *et al.*, 2014). However, through CRISPR/Cas9, researchers have simultaneously edited three homoeoalleles (*TaMLO-A, TaMLO-B, TaMLO-D*) responsible for powdery mildew with a modest mutation rate of 5.6% (Wang *et al.*, 2014). The capability to induce multiple site-specific mutations in polyploid genomes makes this technology suitable for applications in crops like soybean *(Glycine max)*, a paleo-polyploid with mostly two set of genes. In soybean, two genes, i.e. *Glyma01g38150* and *Glyma11g07220*, which are considered orthologues of the *A. thaliana* deficient in DNA methylation 1 gene (DDM1), a gene with a known function as chromatin remodelling factor, have been effectively targeted through sgRNA/Cas9 co-delivery and more than 70% indels frequencies were reported (Jacobs *et al.*, 2015). CRISPR/Cas9-induced mutagenesis has also been reported in two genomic sites of chromosome 4 of soybean, i.e. *DD20* and *DD43*, producing small insertion deletions with 59% and 76% mutation frequencies, respectively (Li *et al.*, 2015). Efficient gene editing has produced mutant loci in maize by precisely targeting upstream of the *liguleless*1 (*LIG1*) gene, *acetolactate synthase* genes *(ALS1, ALS2)* and male fertility genes (*Ms26, Ms45*) (Svitashev *et al.*, 2015).

CRISPR/Cas9 strategies have been successfully used for many agronomically important crop plants such as model *Nicotiana benthamiana* (Li *et al.*, 2013), *Arabidopsis* plants (Li *et al.*, 2013; Jiang *et al.*, 2013), wheat (Wang *et al.*, 2014), maize (Liang *et al.*, 2014), rice (Miao *et al.*, 2013), sorghum (Jiang *et al.*, 2013) and many more (Fig. 4.3 and Table 4.1). Various promoter systems, i.e. *AtU6* or *OsU6* and *AtUBQ, OsUBQ,* or *CaMV 35S*, were exploited to construct sgRNA and Cas9 coding expression vectors for genome editing in these plants (Mao *et al.*, 2013). Several new, quick and novel site-directed mutagenesis protocols are already being established (Shan *et al.*, 2014). About 90% mutation frequency has been calculated in *Arabidopsis* and rice (Miao *et al.*, 2013). The mutation can occur in early embryonic, growth and developmental stages or occur in different part of tissues. The resulting chimera plants contain cells with different genotypes that will be wild type, heterozygous, homozygous or bi-allelic. Different multiple sgRNAs have been employed to create tens to thousands of basepair chromosomal deletions from the genome of *N. benthamiana, Arabidopsis* (Li *et al.*, 2013) and tomato (Brooks *et al.*, 2014). This type of deletion sometimes leads to the loss of an entire gene function (Zhou *et al.*, 2014). Other novel methods include homology-directed repair (HDR) of CRISPR/Cas identified genes, domains, or alleles that were replaced. This method has been successfully used by many researchers along with many important reporter genes such as neomycin transferase (*nptII*) (Li *et al.*, 2013; Feng *et al.*, 2013). Successful insertion of the same gene has been reported into the endogenous *ALS* gene in *Arabidopsis* (Schiml *et al.*, 2014).

In addition to herbaceous plants, the CRISPR/Cas 9 system has been recently utilized in woody plants. Fan *et al.* (2015) reported an improved method of Cas9 gene and multiple sgRNAs delivery to poplar plant (*Populus tricocarpa*) cells to precisely

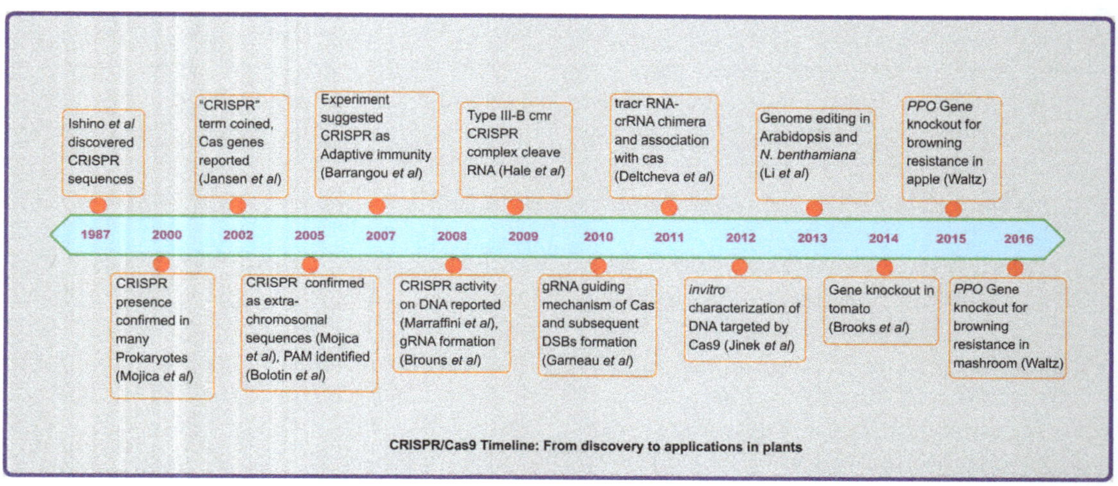

Figure 4.3 CRISPR/Cas9 timeline: from discovery to application in plants.

Table 4.1 CRISPR/Cas9 applications in plants

Species	Trait	Gene Targeted	Mode of action	References
A. thaliana	Multiple traits	*AtPDS3, AtFL2, AtRACK1b, AtRACK1c, AtADH1, AtTT4*	Knock out	Li *et al.*, 2013; Fauser *et al.*, 2014
Oryza sativa	Resistance to bacterial blight	*OsSWEET14; OsSWEET13*	Gene knockout	Zhou *et al.*, 2015a
Zea mays L.	Leaf development; Male fertility; Herbicide resistance	*LIG1; Ms26, Ms45; ALS1, ALS2*	Gene knockout	Liang *et al.*, 2014; Svitashev *et al.*, 2015
S. esculentum	Lignin content; Condensed tannin content	*SlAGO7, Soliyc08g041770, Soliyc07g021170,*	Gene knockout	Brooks *et al.*, 2014
Glycine max L.	Leaf development	*GmDD20, GmDD43, Glyma07g1450, GmDDM1s, GmMIRs*	Gene editing	Li *et al.*, 2015; Jacobs *et al.*, 2015
Triticum aestivum	Powdery mildew resistance, lipoxygenase	*TaMLO-A1, TaLOX2*	Gene editing, knock-out	Wang *et al.*, 2014; Shan *et al.*, 2014
N. benthamiana	Carotenoid biosynthesis	*NbPDS, NbPCNA, NbPDR6*	Gene knockout	Li *et al.*, 2013; Ali *et al.*, 2015; Gao *et al.*, 2015
Citrus sinensis	Transient protein expression in leaves	*CsPDS*	Gene editing	Jia and Wang, 2014
Populus tricocarpa	Leaf development; male fertility; herbicide resistance		Gene knockout	Zhou *et al.*, 2015b
Malus domestica and mushrooms	Browning resistance	*PPO*	Gene knockout	Waltz, 2015, 2016

perturb *phytoene desaturase* gene (*PtoPDS*). Four gRNAs were used to target phytoene desaturase gene 8 (*PtoPDS*). Many albino genotypes were noted after successful genetic transformation via *Agrobacterium*. Maximum mutation frequency of 51.7% was recorded, in which 30 out 59 were homozygous mutants and other two were heterozygous. They concluded that this system can be used for other important woody plants. In sweet orange *Xanthomonas citri,* facilitated agro-infiltration has demonstrated the successful application of cas9/sgRNA system for genome editing (Jia and Wang, 2014).

Harnessing the potential of CRISPR/Cas9 for disease resistance in crops

CRISPR/Cas9 system was successfully used by two research groups to confer long-term resistance against microbes (Ji *et al.*, 2015). These researchers have shown that Cas9/sgRNA inside plant genomes provides durable resistance against Geminivirus in *Nicotiana benthamiana* and *Arabidopsis thaliana* by inhibiting their replication. The direct CRISPR/Cas9 knock-out system was also used in other important crops (Wang *et al.*, 2014). The barley and wheat susceptible *MLO* (Mildew resistance Locus O) gene was targeted to confer long term resistance against powdery mildew disease in these crops (Varallyay *et al.*, 2012). In last 2 years this system has been used to provide broad-spectrum resistance by altering three homeoalleles encoding Mildew Resistance Locus (*MLO*) proteins against fungi in common hexaploid wheat (Wang *et al.*, 2014).

Recently, in soybean a hairy root based CRISPR/Cas9 system was developed where both the sgRNA and Cas9 cassettes were introduced in a single plasmid to enhance transformation frequency (Cai *et al.*, 2015). In a multiplex system targeting 3 genes simultaneously, one of the sgRNAs targeted the *bar* gene and remaining six sgRNA interrupted *GmFEI2* and *GmSHR* genes. Their work showed that the engineered CRISPR/Cas9 system shared maximum efficiency for both endogenous and exogenous genes. A protocol was optimized for delivering truncated gRNA and Cas9 system into *Arabidopsis* to generate new alleles for the *OST2* gene which acts as proton pump in *A. thaliana* (Osakabe *et al.*, 2016). The newly generated mutants with novel alleles exhibited an altered response against abiotic stress in plants. The rapid advancement in CRISPR/Cas9 technology over time is opening new vistas of easy and efficient genome editing in plants with which this technology is posed to revolutionize plants and agricultural research.

Conclusions

The enormous prospect of using CRISPR/Cas9 in crop genome editing and its applications to increase the commercial value of crops are growing by time as the technology is being further developed. We suggest that compared to other genome editing technologies, the relative ease, precision, robustness and efficiency of CRISPR/Cas 9 gives it a unique place to specifically target desirable or undesirable sites for intended genomic response. A vast number of reported loci associated with qualitative and quantitative plant phenotypes like disease resistance, production of phytochemicals and metabolites, important physiological processes and increased yield etc. can be edited via CRISPR/Cas9. The system not only offers means and ways to trigger accurate gene knock-in or knock-outs, but also helps understand complex mechanism of gene expression regulation, epigenetic mechanisms and chromatin remodelling.

However, challenges like off-target mutations, influence of chromatin structure, influence on flanking genes of the target and the effect of delivery methods remain to be investigated in detail. Advanced bioinformatics tools and a large number of Cas9 variants will enable researchers to design gRNA/Cas9 tools requiring longer PAM, which will increase its specificity and reduce off-target effects. Also, with continuous evolution of new pathotypes, fast and timely CRISPR/Cas9 systems for biotic stress resistance in crops should be developed to cope with the pace of pathogens. Despite these issues, as the most efficient tool among gene editing technologies, the research community is enthusiastically improving CRISPR/Cas9 system for its wide applications in plant sciences.

References

Ali, Z., Abul-faraj, A., Li, L., Ghosh, N., Piatek, M., Mahjoub, A., Aouida, M., Piatek, A., Baltes, N.J., Voytas, D.F., *et al.* (2015). Efficient virus-mediated genome editing in plants using the CRISPR/Cas9 system. Mol. Plant.*8*, 1288–1291. http://dx.doi.org/10.1016/j.molp.2015.02.011

Barrangou, R., Fremaux, C., Deveau, H., Richards, M., Boyaval, P., Moineau, S., Romero, D.A., and Horvath, P. (2007). CRISPR provides acquired resistance against viruses in prokaryotes. Science *315*, 1709–1712. http://dx.doi.org/10.1126/science.1138140

Belhaj, K., Chaparro-Garcia, A., Kamoun, S., and Nekrasov, V. (2013). Plant genome editing made easy: targeted mutagenesis in model and crop plants using the CRISPR/Cas system. Plant. Methods *9*, 39. http://dx.doi.org/10.1186/1746-4811-9-39

Bolotin, A., Quinquis, B., Sorokin, A., and Ehrlich, S.D. (2005). Clustered regularly interspaced short palindrome repeats (CRISPRs) have spacers of extrachromosomal origin. Microbiology *151*, 2551–2561. http://dx.doi.org/10.1099/mic.0.28048-0

Bortesi, L., and Fischer, R. (2015). The CRISPR/Cas9 system for plant genome editing and beyond. Biotechnol. Adv. *33*, 41–52. http://dx.doi.org/10.1016/j.biotechadv.2014.12.006

Brooks, C., Nekrasov, V., Lippman, Z.B., and Van Eck, J. (2014). Efficient gene editing in tomato in the first generation using the clustered regularly interspaced short palindromic repeats/CRISPR-associated9 system. Plant Physiol. *166*, 1292–1297. http://dx.doi.org/10.1104/pp.114.247577

Brouns, S.J., Jore, M.M., Lundgren, M., Westra, E.R., Slijkhuis, R.J., Snijders, A.P., Dickman, M.J., Makarova, K.S., Koonin, E.V., and van der Oost, J. (2008). Small CRISPR RNAs guide antiviral defense in prokaryotes. Science *321*, 960–964. http://dx.doi.org/10.1126/science.1159689

Cai, Y., Chen, L., Liu, X., Sun, S., Wu, C., Jiang, B., Han, T., and Hou, W. (2015). CRISPR/Cas9-mediated genome editing in soybean hairy roots. PLOS One *10*, e0136064. http://dx.doi.org/10.1371/journal.pone.0136064

Deltcheva, E., Chylinski, K., Sharma, C.M., Gonzales, K., Chao, Y., Pirzada, Z.A., Eckert, M.R., Vogel, J., and Charpentier, E. (2011). CRISPR RNA maturation by trans-encoded small RNA and host factor RNase III. Nature *471*, 602–607. http://dx.doi.org/10.1038/nature09886

Doudna, J.A., and Charpentier, E. (2014). Genome editing. The new frontier of genome engineering with CRISPR-Cas9. Science *346*, 1258096. http://dx.doi.org/10.1126/science.1258096

Fan, D., Liu, T., Li, C., Jiao, B., Li, S., Hou, Y., and Luo, K. (2015). Efficient CRISPR/Cas9-mediated Targeted Mutagenesis in Populus in the First Generation. Sci. Rep. *5*, 12217. http://dx.doi.org/10.1038/srep12217

Fauser, F., Schiml, S., and Puchta, H. (2014). Both CRISPR/Cas-based nucleases and nickases can be used efficiently for genome engineering in *Arabidopsis* thaliana. Plant. J. *79*, 348–359. http://dx.doi.org/10.1111/tpj.12554

Feng, Z., Zhang, B., Ding, W., Liu, X., Yang, D.L., Wei, P., Cao, F., Zhu, S., Zhang, F., Mao, Y., *et al.* (2013). Efficient genome editing in plants using a CRISPR/Cas system. Cell Res. *23*, 1229–1232. http://dx.doi.org/10.1038/cr.2013.114

Garneau, J.E., Dupuis, M.È., Villion, M., Romero, D.A., Barrangou, R., Boyaval, P., Fremaux, C., Horvath, P., Magadán, A.H., and Moineau, S. (2010). The CRISPR/Cas bacterial immune system cleaves bacteriophage and plasmid DNA. Nature *468*, 67–71. http://dx.doi.org/10.1038/nature09523

Gao, J., Wang, G., Ma, S., Xie, X., Wu, X., Zhang, X., Wu, Y., Zhao, P., and Xia, Q. (2015). CRISPR/Cas9-mediated targeted mutagenesis in Nicotiana tabacum. Plant. Mol. Biol. *87*, 99–110. http://dx.doi.org/10.1007/s11103-014-0263-0

Hale, C.R., Cocozaki, A., Li, H., Terns, R.M., and Terns, M.P. (2014). Target RNA capture and cleavage by the Cmr type III-B CRISPR-Cas effector complex. Genes. Dev. *28*, 2432–2443. http://dx.doi.org/10.1101/gad.250712.114

Hyun, Y., Kim, J., Cho, S.W., Choi, Y., Kim, J.S., and Coupland, G. (2015). Site-directed mutagenesis in *Arabidopsis* thaliana using dividing tissue-targeted RGEN of the CRISPR/Cas system to generate heritable null alleles. Planta *241*, 271–284. http://dx.doi.org/10.1007/s00425-014-2180-5

Ishino, Y., Shinagawa, H., Makino, K., Amemura, M., and Nakata, A. (1987). Nucleotide sequence of the iap gene, responsible for alkaline phosphatase isozyme conversion in *Escherichia coli*, and identification of the gene product. J. Bacteriol. *169*, 5429–5433.

Jacobs, T.B., LaFayette, P.R., Schmitz, R.J., and Parrott, W.A. (2015). Targeted genome modifications in soybean with CRISPR/Cas9. BMC Biotech. *15*, 16. http://dx.doi.org/10.1186/s12896-015-0131-2

Ji, X., Zhang, H., Zhang, Y., Wang, Y., and Gao, C. (2015). Establishing a CRISPR-Cas-like immune system conferring DNA virus resistance in plants. Nat. Plants *1*, 15144. http://dx.doi.org/10.1038/nplants.2015.144

Jia, H., and Wang, N. (2014). Targeted genome editing of sweet orange using Cas9/sgRNA. PLOS One *9*, e93806. http://dx.doi.org/10.1371/journal.pone.0093806

Jiang, W., Zhou, H., Bi, H., Fromm, M., Yang, B., and Weeks, D.P. (2013). Demonstration of CRISPR/Cas9/sgRNA-mediated targeted gene modification in *Arabidopsis*, tobacco, sorghum and rice. Nucleic Acids Res. *41*, e188. http://dx.doi.org/10.1093/nar/gkt780

Jinek, M., Chylinski, K., Fonfara, I., Hauer, M., Doudna, J.A., and Charpentier, E. (2012). A programmable dual-RNA-guided DNA endonuclease in adaptive bacterial immunity. Science *337*, 816–821. http://dx.doi.org/10.1126/science.1225829

Li, J.F., Norville, J.E., Aach, J., McCormack, M., Zhang, D., Bush, J., Church, G.M., and Sheen, J. (2013). Multiplex and homologous recombination-mediated genome editing in *Arabidopsis* and *Nicotiana benthamiana* using guide RNA and Cas9. Nat. Biotechnol. *31*, 688–691. http://dx.doi.org/10.1038/nbt.2654

Li, Z., Liu, Z.B., Xing, A., Moon, B.P., Koellhoffer, J.P., Huang, L., Ward, R.T., Clifton, E., Falco, S.C., and Cigan, A.M. (2015). Cas9-Guide RNA directed genome editing in soybean. Plant Physiol. *169*, 960–970. http://dx.doi.org/10.1104/pp.15.00783

Liang, Z., Zhang, K., Chen, K., and Gao, C. (2014). Targeted mutagenesis in *Zea mays* using TALENs and the CRISPR/Cas system. J. Genet. Genomics *41*, 63–68. http://dx.doi.org/10.1016/j.jgg.2013.12.001

Ma, X., Zhang, Q., Zhu, Q., Liu, W., Chen, Y., Qiu, R., Wang, B., Yang, Z., Li, H., Lin, Y., *et al.* (2015). A robust CRISPR/Cas9 system for convenient, high-efficiency multiplex genome editing in monocot and dicot plants. Mol. Plant. *8*, 1274–1284. http://dx.doi.org/10.1016/j.molp.2015.04.007

Mao, Y., Zhang, H., Xu, N., Zhang, B., Gou, F., and Zhu, J.K. (2013). Application of the CRISPR-Cas system for efficient genome engineering in plants. Mol. Plant. *6*, 2008–2011. http://dx.doi.org/10.1093/mp/sst121

Marraffini, L.A., and Sontheimer, E.J. (2008). CRISPR interference limits horizontal gene transfer in staphylococci by targeting DNA. Science *322*, 1843–1845. http://dx.doi.org/10.1126/science.1165771

Miao, J., Guo, D., Zhang, J., Huang, Q., Qin, G., Zhang, X., Wan, J., Gu, H., and Qu, L.J. (2013). Targeted mutagenesis in rice using CRISPR-Cas system. Cell Res. *23*, 1233–1236. http://dx.doi.org/10.1038/cr.2013.123

Mojica, F.J., García-Martínez, J., and Soria, E. (2005). Intervening sequences of regularly spaced prokaryotic repeats derive from foreign genetic elements.J. Mol. Evol. *60*, 174–182. http://dx.doi.org/10.1007/s00239-004-0046-3

Nekrasov, V., Staskawicz, B., Weigel, D., Jones, J.D., and Kamoun, S. (2013). Targeted mutagenesis in the model plant Nicotiana benthamiana using Cas9 RNA-guided endonuclease. Nat. Biotechnol. *31*, 691–693. http://dx.doi.org/10.1038/nbt.2655

Osakabe, Y., Watanabe, T., Sugano, S.S., Ueta, R., Ishihara, R., Shinozaki, K., and Osakabe, K. (2016). Optimization of CRISPR/Cas9 genome editing to modify abiotic stress responses in plants. Sci. Rep. *6*, 26685. http://dx.doi.org/10.1038/srep26685

Schiml, S., Fauser, F., and Puchta, H. (2014). The CRISPR/Cas system can be used as nuclease for in planta gene targeting and as paired nickases for directed mutagenesis in *Arabidopsis* resulting in heritable progeny. Plant. J. *80*, 1139–1150. http://dx.doi.org/10.1111/tpj.12704

Shan, Q., Wang, Y., Li, J., and Gao, C. (2014). Genome editing in rice and wheat using the CRISPR/Cas system. Nat. Protoc. *9*, 2395–2410. http://dx.doi.org/10.1038/nprot.2014.157

Svitashev, S., Young, J.K., Schwartz, C., Gao, H., Falco, S.C., and Cigan, A.M. (2015). Targeted mutagenesis, precise gene editing, and site-specific gene insertion in maize using Cas9 and guide RNA. Plant Physiol. *169*, 931–945. http://dx.doi.org/10.1104/pp.15.00793

Várallyay, E., Giczey, G., and Burgyán, J. (2012). Virus-induced gene silencing of Mlo genes induces powdery mildew resistance in Triticum aestivum. Arch. Virol. *157*, 1345–1350. http://dx.doi.org/10.1007/s00705-012-1286-y

Waltz, E. (2015). Nonbrowning GM apple cleared for market. Nat. Biotechnol. 33, 326–327. http://dx.doi.org/10.1038/nbt0415-326c

Waltz, E. (2016). Gene-edited CRISPR mushroom escapes US regulation. Nature *532*, 293. http://dx.doi.org/10.1038/nature.2016.19754

Wang, Y., Cheng, X., Shan, Q., Zhang, Y., Liu, J., Gao, C., and Qiu, J.L. (2014). Simultaneous editing of three homoeoalleles in hexaploid bread wheat confers heritable resistance to powdery mildew. Nat. Biotechnol. *32*, 947–951. http://dx.doi.org/10.1038/nbt.2969

Xing, H.L., Dong, L., Wang, Z.P., Zhang, H.Y., Han, C.Y., Liu, B., Wang, X.C., and Chen, Q.J. (2014). A CRISPR/Cas9 toolkit for multiplex genome editing in plants. BMC Plant Biol. *14*, 327. http://dx.doi.org/10.1186/s12870-014-0327-y

Zhang, H., Zhang, J., Wei, P., Zhang, B., Gou, F., Feng, Z., Mao, Y., Yang, L., Zhang, H., Xu, N., *et al.* (2014). The CRISPR/Cas9 system produces specific and homozygous targeted gene editing in rice in one generation. Plant Biotechnol. J. *12*, 797–807. http://dx.doi.org/10.1111/pbi.12200

Zhou, H., Liu, B., Weeks, D.P., Spalding, M.H., and Yang, B. (2014). Large chromosomal deletions and heritable small genetic changes induced by CRISPR/Cas9 in rice. Nucleic Acids Res. *42*,10903-10914. http://dx.doi.org/10.1093/nar/gku806.

Zhou, J., Peng, Z., Long, J., Sosso, D., Liu, B., Eom, J.S., Huang, S., Liu, S., Vera Cruz, C., Frommer, W.B., *et al.* (2015a). Gene targeting by the TAL effector PthXo2 reveals cryptic resistance gene for bacterial blight of rice. Plant. J.*82*, 632–643. http://dx.doi.org/10.1111/tpj.12838

Zhou, X., Jacobs, T.B., Xue, L.J., Harding, S.A., and Tsai, C.J. (2015b). Exploiting SNPs for biallelic CRISPR mutations in the outcrossing woody perennial Populus reveals 4-coumarate:CoA ligase specificity and redundancy. New. Phytol. *208*, 298–301. http://dx.doi.org/10.1111/nph.13470

CRISPR/Cas9-mediated Immunity in Plants Against Pathogens

5

Muhammad Sameeullah[1]*, Faheem Ahmed Khan[2]*, Göksel Özer[3], Noreen Aslam[4], Ekrem Gurel[4], Mohammad Tahir Waheed[5] and Turan Karadeniz[1]

[1]Department of Horticulture, Faculty of Agriculture and Natural Sciences, Abant Izzet Baysal University, Bolu, Turkey.
[2]Key Laboratory of Agricultural Animal Genetics, Breeding and Reproduction, Huazhong Agricultural University, Wuhan, The People's Republic of China.
[3]Department of Plant Protection, Faculty of Agriculture and Natural Sciences, Abant Izzet Baysal University, Bolu, Turkey.
[4]Department of Biology, Faculty of Science and Arts, Abant Izzet Baysal University, Bolu, Turkey.
[5]Department of Biochemistry, Faculty of Biological Sciences, Quaid-I-Azam University, Islamabad, Pakistan.

*Correspondence: sameepbg@gmail.com and faheemgenetics@yahoo.com

https://doi.org/10.21775/9781910190630.05

Abstract

Global crop production is highly threatened due to pathogen invasion. The huge quantity of pesticides application, although harmful to the environment and human health, is carried out to prevent the crop losses worldwide, every year. Therefore, understanding the molecular mechanisms of pathogenicity and plant resistance against pathogens is important. The resistance against pathogens is regulated by three important phytohormones, viz. salicylic acid (SA), jasmonic acid (JA) and ethylene (ET). Here we review the possible role of CRISPR technology to understand the plant pathogenicity by mutating genes responsible for pathogen invasion or up-regulating the phytohormones genes or resistant genes. Thus hormone biosynthesis genes, receptor and feeding genes of pathogens could be important targets for modifications using CRISPR/Cas9 following multiplexing tool box strategy in order to edit multiple genes simultaneously to produce super plants. Here we put forward our idea that the genes would be either mutated in case of plant receptor protein targets of pathogens or up-regulation of resistant genes or hormone biosynthesis genes will be better choice for resistance against pathogens.

Introduction

Immunity

Plants must evolve self-defence systems because they are immobile organisms unable to escape from biotic stress factors such as fungi, viruses and nematodes, and abiotic stress factors like a harsh environment condition. In response to pathogen attack, plants respond using two branched innate immune system (Jones and Dangl, 2006). The function of the first branch is more general to perceive and the response results against many groups of microbes consisting of non-pathogens. The second response is more specific to plant pathogen virulence. Thus, plant immune systems and pathogen elicitors make a complex molecular model across biological kingdoms. Pathogens attack plants to retrieve carbohydrate as a source of energy by penetrating hyphae into plant cell, stylet or through stomata, hydathodes and wounds. First interaction of plant and pathogen establish at the plasma membrane interface to deliver virulence factors into plant cell in order to maintain pathogenicity. In first branch of plant immune system, AMP/pattern-triggered immunity (PTI) and transmembrane pattern recognition receptors (PRRs) respond to elicit microbial or pathogen-associated molecular patterns (MAMPS or PAMPs) (Jones and Dangl, 2006). The second branch of the immune system, effector-triggered immunity (ETI), majorly acts inside the cell using polymorphic NB-LRR protein encoded by R genes (Dangl and Jones, 2001; Marone *et al.*, 2013). The PTI system is a weak and vulnerable immune response system while ETI is associated with strong and robust responses that lead to reactive oxygen species and finally cell death due to hypersensitive response. Typical example of ETI system in which systematic pathway of immune response is triggered upon pathogen attacks (Fig. 5.1). Three plant hormones are associated with plant immune response system, viz. salicylic acid (SA), jasmonic acid (JA) and ethylene (ET). CRISPR/Cas9 has been effectively employed against plant pathogens resistance in plants. Zhou *et al.* (2014) modified a promoter of *OsSWEET* genes for disease resistance, suppression of plant viruses in various plants (Ali *et al.*, 2015a,b; Baltes *et al.*, 2015; Ji *et al.*, 2015) and repression of *MLO* genes target for fungal pathogenicity (Wang *et al.*, 2014). Here we urge on the role of SA, JA and ET defence against pathogens and their regulation. Consequently, CRISPR/Cas9's role in the editing of SA, JA and ET genes mechanism is elaborated in order to explore the target genes which would be of better choice for contributing to plant resistance based on CRISPR/Cas9 system.

SA-, JA- and ET-mediated defence regulation against pathogens

SA is closely associated with immune responses against biotrophic pathogens (Vlot *et al.*, 2009). It has been reported that pathogens like *Ustilagomaydis* Cmu1, *Verticillium dahlia* VdISC1 and *Phytophthorasoja* PsIsc1 inhibited the formation of SA precursors (Djamei *et al.*, 2011; Liu *et al.*, 2014) or the down-regulation of SA signalling pathways (van Damme *et al.*, 2008). Therefore, the detailed binding or suppression of pathogen effectors to SA precursors is essential in order to delete the target binding sites of effectors without interruption of SA biosynthesis by CRISPR/Cas9-mediated genome editing.

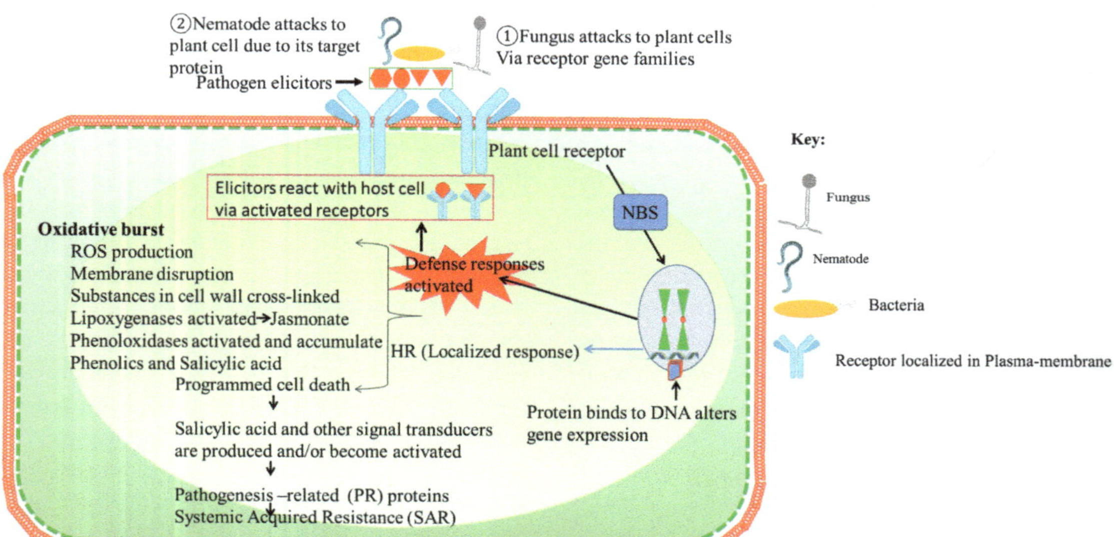

Figure 5.1 Model showing interaction of plant–pathogen at plasma membrane interface of plant cell through the receptors. Series of reactions occurred as a result of pathogen's attack on the plant cell also depicted.

Immunity triggered by JA and its regulation

Jasmonates are plant defence hormones which are lipid-derived molecules and enable plants to adapt under stressful situations due to pathogens and insect invasions. These hormones are produced in response to wounding of tissues and trigger local and long distance defence responses (Koo and Howe, 2009; Larrieu *et al.*, 2015). The defence mechanism triggered due to JA is termed as JA-triggered immunity (JATI). JATI confers broad-spectrum immunity in monocots and dicots. After tissue injury due to pathogens, rapid synthesis of JA and its receptor derivative jasmonoyl-l-isoleucine (JA-Ile) are triggered. Accumulation of JA-Ile in above- and below-ground tissues occurs due to nematode and other pathogens (Fragoso *et al.*, 2014; Grebner *et al.*, 2013; Koo *et al.*, 2009). JA also promotes the proteins and secondary metabolites that have roles in defence including terpenoids, alkaloids and pathogenesis-related proteins (Browse and Howe, 2008; De Geyter *et al.*, 2012; De Vleesschauwer *et al.*, 2013; Gonzales-Vigil *et al.*, 2011). It has been reported in *Arabidopsis* against *Alternaria brassicicola* and *Botrytis cinerea* (Thomma *et al.*, 1998). In addition to *Arabidopsis*, JA-related defence in maize and tomato has also been reported against *Pythium* species (Campos *et al.*, 2014; Staswick *et al.*, 1998; Vijayan *et al.*, 1998; Yan *et al.*, 2012). Constitutive overproductions of JATI regulate the negative effect on plant physiology and development. Since the JA is produced on the expense of limited metabolic resources (Baldwin, 1998; Yan *et al.*, 2007; Zhang and Turner, 2008), the negative feedback mechanism to stop the overproduction or CRISPR/Cas9-mediated pathogen-inducible up-regulation of would be beneficial to lose overburden of plant in terms of consuming metabolites in excess.

Role of ET in plant immunity

When plants are attacked by pathogens, they produce lot of ET (Erb *et al.*, 2012; Lai and Mengiste, 2013; Yang *et al.*, 2013). Few studies showed that the ethylene signalling pathway could contribute to resistance against pathogens (Botanga *et al.*, 2012; Lloyd *et al.*, 2011).

Detailed study of ethylene production due to bacterial pathogen *Pseudomonas syringae* pv. *tomato* DC3000 (Pst) in *Arabidopsis* was reported by Guan *et al.* (2015). The mutants of 1-aminocyclopropane-1-carboxylate synthase (ACS) defective in ethylene induction were highly susceptible to both Pst and Pst-avrRpt2, indicating a potential role of ethylene mediated resistance against bacterial pathogens (Guan *et al.*, 2015). High order mutants of ACS isoforms of *Arabidopsis* were highly susceptible to *Botrytis cinerea* infection compared with wild type, suggesting potential role of ACS genes in pathogen resistance (Tsuchisaka *et al.*, 2009). Two ACS isoforms ACS2 and ACS6 were studied under challenge of *Botrytis cinerea*-induced ethylene biosynthesis and the protein stability was maintained by two *Arabidopsis* pathogen-responsive mitogen-activated protein kinases (MAPKs) MPK3 and MPK6. Later, the reduced level of ethylene induction in the ACS2 and ACS6 mutants indicated that other ACS isoforms play their role in ethylene induction under *Botrytis cinerea* pathogenicity (Han *et al.*, 2010). Therefore, other isoforms of ACS such as ACS7, ACS8 and ACS11 were identified by genetic approach and these isoforms also contribute to pathogen-related ethylene induction. In addition to phosphorylation and protein stability of ACS2 and ACS6, MPK3 and MPK6 also regulate the expression of ACS2 and ACS6 genes by MPK3/MPK6 substrate and WRKY33 (a member of WRKY transcription factor family). Therefore, these findings compel that regulation of ACS genes at transcriptional and post-transcriptional level induce a high level of ethylene production under the challenge of pathogens (Li *et al.*, 2012a), suggesting that ethylene confers resistance against fungal and bacterial pathogens. The manipulation of above mentioned ACS genes by CRISPR/Cas9 following multiplexing can play key role in plant resistance.

CRISPR/Cas9 role in plant immunity

CRISPR/Cas9 defence mechanism against invasions has been exploited to use in biological disciplines to knock-out genes or enhance the expression of target genes. There are three types of CRISPR/Cas9; however, Type II gains more attention for its utility and versatility for genome editing (Bhaya *et al.*, 2011). In a typical example of CRISPR/Cas9 system, Cas9 enzyme has two lobes REC and NUC which are activated by subgenomic RNA (sgRNA) to develop Cas9–sgRNA complex. After complex formation, Cas9 searches for protospacer adjacent motif (PAM). Once PAM is determined, initial DNA destabilization and distortion occurs which leads to DNA cleavage in a systematic order (Belhaj *et al.*, 2015). Then Cas9, a sequence-specific nuclease, creates double-strand breaks (DSB) in the target region of DNA. These DSB are repaired by homology directed repair (HDR) or non-homologous end joining (NHEJ) that is a cell's endogenous repairing system. There are an increasing number of reports showing NHEJ repair in comparison to HDR to repair DSB in CRISPR/Cas9 genome editing. Based on the available literature, this trend has been drawn and presented in Fig. 5.2. With the advancement of transcriptomics of plants invaded by pathogens are available publicly such as NCBI gene expression omnibus (GEO) or at Array express. While utilizing KEGG pathways to understand the disease susceptible and resistant gene families for a particular pathogen under specific conditions, CRISPR/Cas has potential to address the functional characterization of gene families induced under pathogenicity in no time. On the other hand, due to the short life cycle of pathogens, they have evolved resistance against the plant native immune system by silencing host RNA and making it unsuitable for crop resistance (Pumplin and Voinnet, 2013). In order to overcome this bottleneck, a more robust strategy is to utilize gene targeting (GT), which permits disruption of endogenous or addition of *in vitro* manipulated copy of gene. Although this GT technology was a powerful tool to edit genes in plants for plant protection, it was limited due to its rare double-strand breaks (DSBs). However, this potential barrier has been overcome by the development of TALENS and CRISPR/Cas9 system by utilization of novel endonucleases (Puchta and Fauser, 2013).

Up-regulation of target genes by CRISPR (dCas9-VP64)

Transcriptional activation mediated by CRISPR was reported in humans at a genome-wide scale in a multiplex gene fashion (Konermann *et al.*, 2014). This was based on deactivated Cas9 (dCas9) mediated transcriptional system to enhance the target genes expression. The first tool box in plants was transcriptional activator VP64 fused with dCas9 that was reported to enhance the multiple endogenous genes expression in *Arabidopsis* (*Arabidopsis thaliana*), rice (*Oryza sativa*) and tobacco (*Nicotiana benthamiana*) (Lowder *et al.*, 2015). In this study, it was suggested that the modification in promoter regions of endogenous genes with fused complex of dCas9-VP64 could enhance the gene expression in transient or stable transformation. The expression of the genes could enhance 2- to 7-fold in the case of AtPAP1 and 3- to 7.5-fold of miR319 as compared to control in *Arabidopsis*. The GUS reporter gene expression mediated by dCas9-VP64 enhanced remarkably compared to control. Thus this study strongly suggests the successful activation of transcriptional regulation of the endogenous genes by

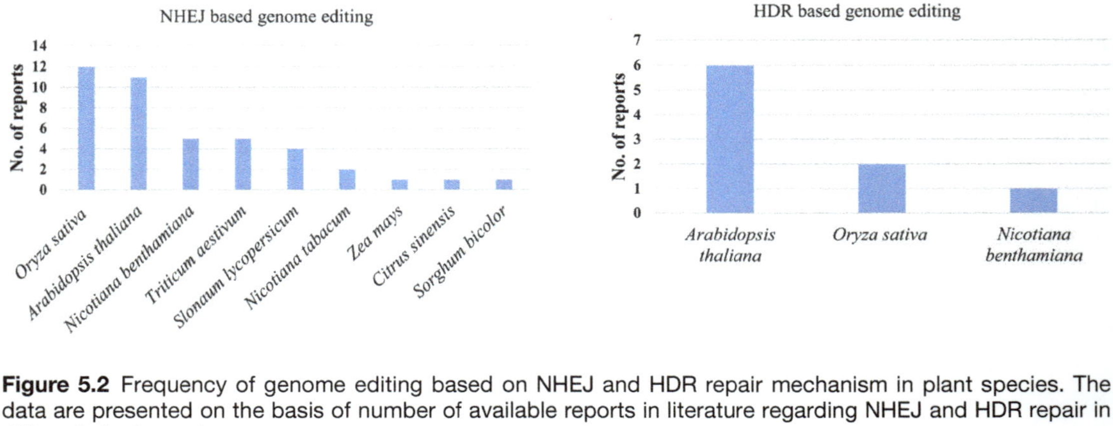

Figure 5.2 Frequency of genome editing based on NHEJ and HDR repair mechanism in plant species. The data are presented on the basis of number of available reports in literature regarding NHEJ and HDR repair in different plant species.

dCas9-VP64 system. Therefore, it will be feasible to adopt this up-regulation of transcriptional system in order to enhance the expression level of target genes related to pathogens resistance.

Multiplexing of genome editing

The editing of multiple genes is of great interest in plant immune systems since several gene family members could be contributing to pathogenicity responses. Multiplex gene editing in plants was reported by Li *et al.* (2013) in *Arabidopsis* and tobacco by editing *AtPDS3* or *NbPDS* target sgRNA by a single binary plasmid containing pcoCas9 in order to suppress carotenoid biosynthesis and chlorophyll oxidation with expected photobleached phenotype.

Li *et al.* (2012b) demonstrated TALEN based editing of *OsSWEET14* gene which is target of rice bacterial blight susceptibility. This gene encodes SWEET efflux transporter and manipulated by *X. oryzae* pv. *oryzae* while using TAL effectors AvrXa7 or PthXo3 to induce the gene expression and leading to stealing sugar from the plant cell in order to feed the pathogen for its growth. The successful editing of *OsSWEET14* by TALEN could produce disease resistant rice plants. This group also demonstrated the successful site-specific mutation in the promoter regions of disease susceptible genes *SWEET11* and *SWEET14* (Li *et al.*, 2012b) based on CRISPR/Cas9 system (Jiang *et al.*, 2013; Zhou *et al.*, 2014) where four SWEET transporters *OsSWEET1a*, *OsSWEET1b*, *OsSWEET11*, *OsSWEET13* were mutated successfully. The simultaneous editing of same gene family members was edited at higher efficiency (87–100%). In this study merely establishment of efficient multiplexing editing target for disease susceptible target genes. However, no pathogenicity test was performed to show the efficacy at plant pathogen interaction. The multiple sgRNA members of a gene family expressed by U3 and U6 small nuclear RNA promoters from rice: OsU3, OsU6b, OsU6c (Shan *et al.*, 2013a; Xie *et al.*, 2015; Zhou *et al.*, 2014), and OsU6a (Ma *et al.*, 2015). Therefore, multiplexing would facilitate in editing targets for several pathogens simultaneously by targeting multiple genes which could serve as target of pathogenicity.

Resistance against fungal pathogenicity

Botrytis cinerea (grey mould) is an important fruit rot fungus which reduces shelf life of fleshy fruits and berries and also causes leaf blight. Five resistance and two susceptible quantitative trait loci (QTLs) for grey mould were identified in wild tomato (Davis *et al.*, 2009). Three grey mould resistance locus (GML) and sensitive QTL shave been reported for tomato. Inactivation of sensitive gene or transcriptional activation of resistant gene would reduce the grey mould pathogenicity and ensure the crop safety thus increasing the yield potential and reduction of post-harvest losses. SWEET effluxers are target for pathogens/symbionts to acquire energy as a source of carbohydrate (Li *et al.*, 2012b; Sameeullah *et al.*, 2016). Modification in the promoter region of *OsSWEET* could enhance disease resistance (Zhou *et al.*, 2014). Powdery mildew devastates crop of cereals worldwide every year. The target of powdery mildew in plant cells is the MLO gene family which was first discovered in barley (Büschges *et al.*, 1997). Later on the homologues were identified in other plant species as well. Recently, CRISPR/Cas9 based edited MLO genes in wheat (*Triticum aestivum*) were reported to knock-out the target of powdery mildew in order to create resistant wheat (Wang *et al.*, 2014). The effector Avr4/6 of an oomycete pathogen *Phytophthora sojae*, causing root and stem rot of soybean, was effectively knocked out revealed the corresponding R gene loci RPS4 and RPS6 (Fang and Tyler, 2016). Ethylene response factor (ERF) gene *OsERF922* which is actively induced by multiple stress factors or plant hormones such as salt and rice blast causal agent pushovers of *Magnaporthe oryzae* and ABA. Enhanced resistance against *M. oryzae* was observed in RNAi based knockdown rice plants. On the other hand overexpression of the *OsERF922* gene made plants vulnerable and susceptible to *M. oryzae* (Liu *et al.*, 2012). Recently, Wang *et al.* (2016) demonstrated targeted mutagenesis in *OsERF922* by targeting multiple sites while using Cas9-multi-target-sgRNAs (C-ERF922S1S2 and C-ERF922S1S2S3) in order to obtain plants containing mutations at multiple sites. Edited rice plants showed a decreased level of pathogenicity compared to wild type at seedling and tillering stages. Moreover, there was no significant difference among agronomic traits of mutant lines and wild type. Therefore, CRISPR/Cas9 presents a sustainable approach in plant protection against pathogens without disturbing genetic yield potential of commercial cultivars. This powerful tool can be applied to other high yielding elite cultivars of crop plants which has been obsolete due to invasion of pathogens. In this context there is an example of cotton cultivar NIAB-78 which had high yielding potential but was superseded due to cotton leaf curl virus disease (CLCuD).

Resistance against nematode infestation

Nematodes have feeding sites in plant cells to infect the plant cell to get energy. Inactivation of genes that are induced during nematode infestation would reduce the nematode infestation. Genes which are induced upon nematode infestation have been reported in many plants (Gheysen and Fenoll, 2002; Kyndt *et al.*, 2012). The power of CRISPR/Cas9 based genome editing would be a possible solution to repress the induced genes by following multiplexing tool box (Lowder *et al.*, 2015). Recently, novel target genes have been identified that could play role in establishment of parasitism between root knot nematode and the plant, especially plant nutrient transporters (Kyndt *et al.*, 2012). Inactivation of those genes by RNAi significantly reduced the degree of parasitism (Danchin *et al.*, 2013). Thus, CRISPR/Cas9 can be a promising strategy to inactivate multiple target genes simultaneously and observe the level of parasitism in host plants in no time.

Resistance against plant viruses

Plant viruses are the notorious pathogens of the plant kingdom. CRISPR/Cas9 has been effectively exploited to reduce the virulence of viruses in different crop plants against geminiviruses (Chaparro-Garcia *et al.*, 2015). sgRNA of tomato yellow leaf curl virus (TYLCV) were introduced into *Nicotiana benthamiana* that could effectively reduce the viral infection (Ali *et al.*, 2015b). Bean yellow dwarf virus (BeYDV) genome was edited by CRISPR/Cas9 and effectively reduced the copy number of the virus, thus reducing the infestation rate and conferred tolerance against virus (Baltes *et al.*, 2015). Ji *et al.* (2015) demonstrated induction of mutation and inhibition of beet severe curly top virus (BSCTV) by introducing sgRNA into wild tobacco. Their findings showed that overexpression in *Arabidopsis* and *N. benthamiana* conferred high resistance to virus infection.

Conclusion

Plant immunity is a rapidly growing subject to understand plant–pathogen interactions at a molecular level by genomic, transcriptomic, proteomic and metabolomic approaches. However, little is known so far about jasmonic acid and ethylene hormone impact in plant immunity. CRISPR/Cas9 tool is a hotcake tool for genome editing in crop plants for various traits of interest. CRISPR/Cas9 tool can knock-out or up-regulate the expression of target genes to confer tolerance and resistance against pathogens. In the latter case, manipulation of promoter of target genes is required. The multifaceted of CRISPR/Cas9 to manipulate plant resistant genes or susceptible genes to intervene resistance against multiple pathogens. A model is shown in (Fig. 5.3). The use of this highly precise tool can increase protection of crops. Additionally, post-harvest losses may also be reduced due to decrease in pathogens invasions especially in fleshy fruits. The application of this technology will not only boost crop yield but also ensure the acceptance of commodity to the consumer end.

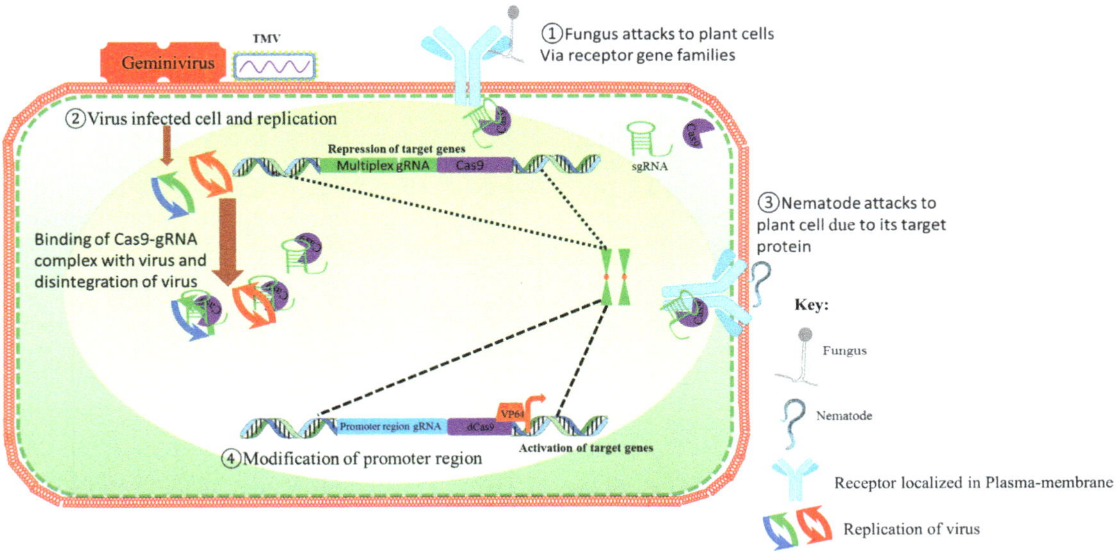

Figure 5.3 Model to show the repression or activation of sensitive or resistant genes in plant cell to confer resistance against the fungi, virus and nematode pathogens. Following multiplexing strategy of CRISPR/Cas9, genome editing would inactivate the targets of pathogens in a plant cell. Transcriptional activation of target gene by modification in promoter region mediated by dCas9 would enhance the expression level of resistant genes that will confer resistance against pathogens. (1) Shows the fungal interaction with plant receptors which are targets for fungal pathogenicity into plant cell. Inhibition of the target receptors by CRISPR/Cas9 localized at the plasma-membrane would induce plant resistance against the pathogens. (2) Attack of geminivirus or tobacco mosaic virus (TMV) into plant cell and its replication. Binding of target sgRNA to plant genome disintegrates the viral genome and thus induces the resistance against viruses in plant cell. (3) Nematode infection at plant cell interface induced plant nutrient transporters localized at the plasma membrane and thus facilitates transport of nutrients to nematode. Thus, targeting those nutrient transporters by CRISPR/Cas9 multiplexing tool to silence the target transporters would facilitate resistance against nematode infection. (4) Modification in promoter regions of target genes which on mutation could affect plant physiological development therefore, over-expression by modification promoter region mediated by CRISPR/Cas9 system contributes to plant resistance against pathogens. In this context promotion of genes like SA, JA and ET biosynthesis would be target to enhance expression in order to confer resistance against pathogens.

References

Ali, Z., Abul-faraj, A., Li, L., Ghosh, N., Piatek, M., Mahjoub, A., Aouida, M., Piatek, A., Baltes, Nicholas J., Voytas, Daniel F., Dinesh-Kumar, S., Mahfouz, Magdy M. (2015a). Efficient Virus-Mediated Genome Editing in Plants Using the CRISPR/Cas9 System. Mol. Plant. *8*, 1288–1291.

Ali, Z., Abulfaraj, A., Idris, A., Ali, S., Tashkandi, M., and Mahfouz, M.M. (2015b). CRISPR/Cas9-mediated viral interference in plants. Genome Biol. *16*, 238. http://dx.doi.org/10.1186/s13059-015-0799-6

Baldwin, I.T. (1998). Jasmonate-induced responses are costly but benefit plants under attack in native populations. Proc. Natl. Acad. Sci. U.S.A. *95*, 8113–8118.

Baltes, N.J., Hummel, A.W., Konecna, E., Cegan, R., Bruns, A.N., Bisaro, D.M., and Voytas, D.F. (2015). Conferring resistance to geminiviruses with the CRISPR–Cas prokaryotic immune system. Nature Plants. *1*, 15145.

Belhaj, K., Chaparro-Garcia, A., Kamoun, S., Patron, N.J., and Nekrasov, V. (2015). Editing plant genomes with CRISPR/Cas9. Curr. Opin. Biotechnol. *32*, 76–84. http://dx.doi.org/10.1016/j.copbio.2014.11.007

Bhaya, D., Davison, M., and Barrangou, R. (2011). CRISPR-Cas systems in bacteria and archaea: versatile small RNAs for adaptive defense and regulation. Annu. Rev. Genet. *45*, 273–297. http://dx.doi.org/10.1146/annurev-genet-110410-132430

Botanga, C.J., Bethke, G., Chen, Z., Gallie, D.R., Fiehn, O., and Glazebrook, J. (2012). Metabolite profiling of *Arabidopsis* inoculated with Alternaria brassicicola reveals that ascorbate reduces disease severity. Mol. Plant-Microbe Interact.*25*, 1628–1638.

Browse, J., and Howe, G.A. (2008). New weapons and a rapid response against insect attack. Plant Physiol. *146*, 832–838. http://dx.doi.org/10.1104/pp.107.115683

Büschges, R., Hollricher, K., Panstruga, R., Simons, G., Wolter, M., Frijters, A., van Daelen, R., van der Lee, T., Diergaarde, P., Groenendijk, J., *et al.* (1997). The barley Mlo gene: a novel control element of plant pathogen resistance. Cell *88*, 695–705.

Campos, M.L., Kang, J.H., and Howe, G.A. (2014). Jasmonate-triggered plant immunity. J. Chem. Ecol. *40*, 657–675. http://dx.doi.org/10.1007/s10886-014-0468-3

Chaparro-Garcia, A., Kamoun, S., and Nekrasov, V. (2015). Boosting plant immunity with CRISPR/Cas. Genome Biol. *16*, 254. http://dx.doi.org/10.1186/s13059-015-0829-4

Danchin, E.G., Arguel, M.J., Campan-Fournier, A., Perfus-Barbeoch, L., Magliano, M., Rosso, M.N., Da Rocha, M., Da Silva, C., Nottet, N., Labadie, K., *et al.* (2013). Identification of novel target genes for safer and more specific control of root-knot nematodes from a pan-genome mining. PLOS Pathog. *9*, e1003745. http://dx.doi.org/10.1371/journal.ppat.1003745

Dangl, J.L., and Jones, J.D. (2001). Plant pathogens and integrated defence responses to infection. Nature *411*, 826–833. http://dx.doi.org/10.1038/35081161

Davis, J., Yu, D., Evans, W., Gokirmak, T., Chetelat, R.T., and Stotz, H.U. (2009). Mapping of loci from Solanum lycopersicoides conferring resistance or susceptibility to Botrytis cinerea in tomato. Theor. Appl. Genet. *119*, 305–314.

De Geyter, N., Gholami, A., Goormachtig, S., and Goossens, A. (2012). Transcriptional machineries in jasmonate-elicited plant secondary metabolism. Trends Plant Sci. *17*, 349–359. http://dx.doi.org/10.1016/j.tplants.2012.03.001

De Vleesschauwer, D., Gheysen, G., and Höfte, M. (2013). Hormone defense networking in rice: tales from a different world. Trends Plant Sci. *18*, 555–565. http://dx.doi.org/10.1016/j.tplants.2013.07.002

Djamei, A., Schipper, K., Rabe, F., Ghosh, A., Vincon, V., Kahnt, J., Osorio, S., Tohge, T., Fernie, A.R., Feussner, I., *et al.* (2011). Metabolic priming by a secreted fungal effector. Nature *478*, 395–398. http://dx.doi.org/10.1038/nature10454

Erb, M., Meldau, S., and Howe, G.A. (2012). Role of phytohormones in insect-specific plant reactions. Trends Plant Sci. *17*, 250–259. http://dx.doi.org/10.1016/j.tplants.2012.01.003

Fragoso, V., Rothe, E., Baldwin, I.T., and Kim, S.G. (2014). Root jasmonic acid synthesis and perception regulate folivore-induced shoot metabolites and increase Nicotiana attenuata resistance. New. Phytol. *202*, 1335–1345. http://dx.doi.org/10.1111/nph.12747

Gheysen, G., and Fenoll, C. (2002). Gene expression in nematode feeding sites. Annu. Rev. Phytopathol. *40*, 191–219. http://dx.doi.org/10.1146/annurev.phyto.40.121201.093719

Gonzales-Vigil, E., Bianchetti, C.M., Phillips, G.N., and Howe, G.A. (2011). Adaptive evolution of threonine deaminase in plant defense against insect herbivores. Proc. Natl. Acad. Sci. U.S.A. *108*, 5897–5902. http://dx.doi.org/10.1073/pnas.1016157108

Grebner, W., Stingl, N.E., Oenel, A., Mueller, M.J., and Berger, S. (2013). Lipoxygenase6-dependent oxylipin synthesis in roots is required for abiotic and biotic stress resistance of *Arabidopsis*. Plant Physiol. *161*, 2159–2170. http://dx.doi.org/10.1104/pp.113.214544

Guan, R., Su, J., Meng, X., Li, S., Liu, Y., Xu, J., and Zhang, S. (2015). Multilayered regulation of ethylene induction plays a positive role in *Arabidopsis* resistance against *Pseudomonas syringae*. Plant Physiol. *169*, 299–312.

Han, L., Li, G.J., Yang, K.Y., Mao, G., Wang, R., Liu, Y., and Zhang, S. (2010). Mitogen-activated protein kinase 3 and 6 regulate Botrytis cinerea-induced ethylene production in *Arabidopsis*. Plant J. *64*, 114–127.

Ji, X., Zhang, H., Zhang, Y., Wang, Y., and Gao, C. (2015). Establishing a CRISPR-Cas-like immune system conferring DNA virus resistance in plants. Nat. Plants *1*, 15144. http://dx.doi.org/10.1038/nplants.2015.144

Jiang, W., Zhou, H., Bi, H., Fromm, M., Yang, B., and Weeks, D.P. (2013). Demonstration of CRISPR/Cas9/sgRNA-mediated targeted gene modification in *Arabidopsis*, tobacco, sorghum and rice. Nucleic Acids Res. *41*, e188. http://dx.doi.org/10.1093/nar/gkt780

Jones, J.D., and Dangl, J.L. (2006). The plant immune system. Nature *444*, 323–329.

Konermann, S., Brigham, M.D., Trevino, A.E., Joung, J., Abudayyeh, O.O., Barcena, C., Hsu, P.D., Habib, N., Gootenberg, J.S., and Nishimasu, H. (2014). Genome-scale transcriptional activation by an engineered CRISPR-Cas9 complex. Nature *517*, 583–588.

Koo, A.J., Gao, X., Jones, A.D., and Howe, G.A. (2009). A rapid wound signal activates the systemic synthesis of bioactive jasmonates in *Arabidopsis*. Plant. J. *59*, 974–986. http://dx.doi.org/10.1111/j.1365-313X.2009.03924.x

Koo, A.J., and Howe, G.A. (2009). The wound hormone jasmonate. Phytochemistry *70*, 1571–1580. http://dx.doi.org/10.1016/j.phytochem.2009.07.018

Kyndt, T., Denil, S., Haegeman, A., Trooskens, G., Bauters, L., Van Criekinge, W., De Meyer, T., and Gheysen, G. (2012). Transcriptional reprogramming by root knot and migratory nematode infection in rice. New. Phytol. *196*, 887–900. http://dx.doi.org/10.1111/j.1469-8137.2012.04311.x

Lai, Z., and Mengiste, T. (2013). Genetic and cellular mechanisms regulating plant responses to necrotrophic pathogens. Curr. Opin. Plant Biol. *16*, 505–512. http://dx.doi.org/10.1016/j.pbi.2013.06.014

Larrieu, A., Champion, A., Legrand, J., Lavenus, J., Mast, D., Brunoud, G., Oh, J., Guyomarc'h, S., Pizot, M., Farmer, E.E., Turnbull, C., Vernoux, T., Bennett, M.J., and Laplaze, L. (2015). A fluorescent hormone biosensor reveals the dynamics of jasmonate signalling in plants. Nat. Commun. *6*. http://dx.doi.org/10.1038/ncomms7043

Li, G., Meng, X., Wang, R., Mao, G., Han, L., Liu, Y., and Zhang, S. (2012a). Dual-level regulation of ACC synthase activity by MPK3/MPK6 cascade and its downstream WRKY transcription factor during ethylene induction in *Arabidopsis*. PLOS Genet. *8*, e1002767. http://dx.doi.org/10.1371/journal.pgen.1002767

Li, T., Liu, B., Spalding, M.H., Weeks, D.P., and Yang, B. (2012b). High-efficiency TALEN-based gene editing produces disease-resistant rice. Nat. Biotechnol. *30*, 390–392. http://dx.doi.org/10.1038/nbt.2199

Li, J.-F., Norville, J.E., Aach, J., McCormack, M., Zhang, D., Bush, J., Church, G.M., and Sheen, J. (2013). Multiplex and homologous recombination-mediated genome editing in *Arabidopsis* and *Nicotiana benthamiana* using guide RNA and Cas9. Nat. Biotechnol.. *31*, 688–691.

Liu, D., Chen, X., Liu, J., Ye, J., and Guo, Z. (2012). The rice ERF transcription factor OsERF922 negatively regulates resistance to Magnaporthe oryzae and salt tolerance. J. Exp. Bot. *63*, 3899–3911. http://dx.doi.org/10.1093/jxb/ers079

Liu, T., Song, T., Zhang, X., Yuan, H., Su, L., Li, W., Xu, J., Liu, S., Chen, L., Chen, T., *et al.* (2014). Unconventionally secreted effectors of two filamentous pathogens target plant salicylate biosynthesis. Nat. Commun. *5*, 4686. http://dx.doi.org/10.1038/ncomms5686

Lloyd, A.J., William Allwood, J., Winder, C.L., Dunn, W.B., Heald, J.K., Cristescu, S.M., Sivakumaran, A., Harren, F.J., Mulema, J., Denby, K., Goodacre, R., Smith, A.R., and Mur, L.A. (2011). Metabolomic approaches reveal that cell wall modifications play a major role in ethylene-mediated resistance against *Botrytis cinerea*. Plant J. *67*, 852–868.

Lowder, L.G., Zhang, D., Baltes, N.J., Paul, J.W., Tang, X., Zheng, X., Voytas, D.F., Hsieh, T.F., Zhang, Y., and Qi, Y. (2015). A CRISPR/Cas9 toolbox for multiplexed plant genome editing and transcriptional regulation. Plant Physiol. *169*, 971–985. http://dx.doi.org/10.1104/pp.15.00636

Ma, X., Zhang, Q., Zhu, Q., Liu, W., Chen, Y., Qiu, R., Wang, B., Yang, Z., Li, H., Lin, Y., *et al.* (2015). A robust CRISPR/Cas9 system for convenient, high-efficiency multiplex genome editing in monocot and dicot plants. Mol. Plant. *8*, 1274–1284. http://dx.doi.org/10.1016/j.molp.2015.04.007

Marone, D., Russo, M.A., Laidò, G., De Leonardis, A.M., and Mastrangelo, A.M. (2013). Plant nucleotide binding site–leucine-rich repeat (nbs-lrr) genes: active guardians in host defense responses. Int. J. Mol. Sci.. *14*, 7302–7326.

Puchta, H., and Fauser, F. (2013). Gene targeting in plants: 25 years later. Int. J. Dev. Biol. *57*, 629–637. http://dx.doi.org/10.1387/ijdb.130194hp

Pumplin, N., and Voinnet, O. (2013). RNA silencing suppression by plant pathogens: defence, counter-defence and counter-counter-defence. Nat. Rev. Microbiol. *11*, 745–760. http://dx.doi.org/10.1038/nrmicro3120

Sameeullah, M., Demiral, T., Aslam, N., Baloch, F.S., and Gurel, E. (2016). In silico functional analyses of SWEETs reveal cues for their role in amf symbiosis. In: Plant, Soil and Microbes: Volume 2: Mechanisms and Molecular Interactions (R.K. Hakeem, S.M. Akhtar, eds.) pp. 45–58, Springer International Publishing, Cham.

Staswick, P.E., Yuen, G.Y., and Lehman, C.C. (1998). Jasmonate signaling mutants of *Arabidopsis* are susceptible to the soil fungus *Pythium irregulare*. Plant. J. *15*, 747–754.

Thomma, B.P., Eggermont, K., Penninckx, I.A., Mauch-Mani, B., Vogelsang, R., Cammue, B.P., and Broekaert, W.F. (1998). Separate jasmonate-dependent and salicylate-dependent defense-response pathways in *Arabidopsis* are essential for resistance to distinct microbial pathogens. Proc. Natl. Acad. Sci. U.S.A. *95*, 15107–15111.

Tsuchisaka, A., Yu, G., Jin, H., Alonso, J.M., Ecker, J.R., Zhang, X., Gao, S., and Theologis, A. (2009). A combinatorial interplay among the 1-aminocyclopropane-1-carboxylate isoforms regulates ethylene biosynthesis in *Arabidopsis thaliana*. Genetics. *183*, 979–1003.

van Damme, M., Huibers, R.P., Elberse, J., and Van den Ackerveken, G. (2008). *Arabidopsis* DMR6 encodes a putative 2OG-Fe(II) oxygenase that is defense-associated but required for susceptibility to downy mildew. Plant. J. *54*, 785–793. http://dx.doi.org/10.1111/j.1365-313X.2008.03427.x

Vijayan, P., Shockey, J., Lévesque, C.A., Cook, R.J., and Browse, J. (1998). A role for jasmonate in pathogen defense of *Arabidopsis*. Proc. Natl. Acad. Sci. U.S.A. *95*, 7209–7214.

Vlot, A.C., Dempsey, D.A., and Klessig, D.F. (2009). Salicylic Acid, a multifaceted hormone to combat disease. Annu. Rev. Phytopathol. *47*, 177–206. http://dx.doi.org/10.1146/annurev.phyto.050908.135202

Wang, F., Wang, C., Liu, P., Lei, C., Hao, W., Gao, Y., Liu, Y.G., and Zhao, K. (2016). Enhanced rice blast resistance by CRISPR/Cas9-targeted mutagenesis of the ERF transcription factor gene OsERF922. PLOS One *11*, e0154027. http://dx.doi.org/10.1371/journal.pone.0154027

Wang, Y., Cheng, X., Shan, Q., Zhang, Y., Liu, J., Gao, C., and Qiu, J.L. (2014). Simultaneous editing of three homoeoalleles in hexaploid bread wheat confers heritable resistance to powdery mildew. Nat. Biotechnol. *32*, 947–951. http://dx.doi.org/10.1038/nbt.2969

Xie, K., Minkenberg, B., and Yang, Y. (2015). Boosting CRISPR/Cas9 multiplex editing capability with the endogenous tRNA-processing system. Proc. Natl. Acad. Sci. U.S.A. *112*, 3570–3575. http://dx.doi.org/10.1073/pnas.1420294112

Yan, Y., Christensen, S., Isakeit, T., Engelberth, J., Meeley, R., Hayward, A., Emery, R.J., and Kolomiets, M.V. (2012). Disruption of OPR7 and OPR8 reveals the versatile functions of jasmonic acid in maize development and defense. Plant. Cell. *24*, 1420–1436. http://dx.doi.org/10.1105/tpc.111.094151

Yan, Y., Stolz, S., Chételat, A., Reymond, P., Pagni, M., Dubugnon, L., and Farmer, E.E. (2007). A downstream mediator in the growth repression limb of the jasmonate pathway. Plant. Cell. *19*, 2470–2483.

Yang, D.L., Yang, Y., and He, Z. (2013). Roles of plant hormones and their interplay in rice immunity. Mol. Plant. *6*, 675–685. http://dx.doi.org/10.1093/mp/sst056

Zhang, Y., and Turner, J.G. (2008). Wound-induced endogenous jasmonates stunt plant growth by inhibiting mitosis. PLOS One 3, e3699. http://dx.doi.org/10.1371/journal.pone.0003699

Zhou, H., Liu, B., Weeks, D.P., Spalding, M.H., and Yang, B. (2014). Large chromosomal deletions and heritable small genetic changes induced by CRISPR/Cas9 in rice. Nucleic Acids Res. *42*, 10903–10914. http://dx.doi.org/10.1093/nar/gku806

Improving CRISPR/Cas9 On-target Specificity

6

Muhammad Jamal[1*], Arif Ullah[2], Muhammad Ahsan Naeem[1], Rohit Tyagi[1], Zeshan Habib[1] and Khaista Rehman[1]

[1]State Key Laboratory of Agricultural Microbiology, College of Animal Science and Medicine, Huazhong Agricultural University, Wuhan, The People's Republic of China.
[2]Institute of Biochemistry and Biotechnology, University of Veterinary and Animal Sciences, Lahore, Pakistan.

*Corresponding author: jamalbiotech@webmail.hzau.edu.cn

https://doi.org/10.21775/9781910190630.06

Abstract

The CRISPR/Cas9 has revolutionized the field of molecular biology, medical genetics and medicine. The technology is robust, facile and simple to achieve genome targeting in cells and organisms. However, to propagate these nucleases for therapeutic application, the on-target specificity is of paramount importance. Although the binding and cleavage of off-target sites by Cas9 is an issue of concern, however, the specificity of CRISPR technology is greatly improved in current research employing the use of engineer nucleases, improved gRNA selection, novel Cas9 orthologues and the advancement in methods to detect and screen off-target sites and its effects. Here we summarize the advances in this state-of-the-art technology that will equip the genome editing tools to be applied in clinical research. The researcher should optimize these methods with emphasis to achieving perfection in the specificity.

Introduction

CRISPR (clustered, regularly interspaced, short palindromic repeat)/Cas (CRISPR-associated protein) is a defensive mechanism against the invading pathogen such as phages and plasmids (Bhaya *et al.*, 2011; Jinek *et al.*, 2012; Wiedenheft *et al.*, 2012). The system works by integrating a part of the invading genome into the specialized region in the genome known as repeats; the incorporated part known as a spacer is inserted in between the two repeats. The expression of the repeat-spacer-repeat produces pre-CrRNA which upon maturation liberates mature crRNA. The degradation of foreign genome is achieved by means of CrRNA (CRISPR RNAs) complementary to sequences within re-invading viral genome or plasmid DNA (Bhaya *et al.*, 2011; Terns and Terns, 2011; Wiedenheft *et al.*, 2012). The cRNA guides Cas9 to the target site on the DNA to produce breaks on both DNA strands thereby removing the foreign genome (Jinek *et al.*, 2012). So far, CRISPR/Cas systems are recognized in a representative set of 703 archaeal and bacterial genomes, of which 310 (44%) encode one or more CRISPR/Cas modules after its initial discovery in *E. coli* in 1987 (Deveau *et al.*, 2010; Koonin and Makarova, 2009).

CRISPR technology has been established by the researchers to regulate or edit any locus of interest. Cas9 is localized by means by a short stretch of guide sequence known as guide RNA (gRNA). From the time of its first application more than 1700 papers have been published and more than 60,000 CRISPR/Cas related plasmids have been shipped by Addgene. The tool has been successfully applied in many organisms for genome editing including *C. elegans* (Friedland *et al.*, 2013; Tzur *et al.*, 2013), zebrafish (Chang *et al.*, 2013), mice (Platt *et al.*, 2014), rats (Li *et al.*, 2013), rabbits

(Yang *et al.*, 2014) and monkeys (Niu *et al.*, 2014), and also applied in cell lines and to treat mutations causing genetic disorders (Khan *et al.*, 2016; Niu *et al.*, 2014) such as β-thalassaemia (Xie *et al.*, 2014), cystic fibrosis (Schwank *et al.*, 2013), haemophilia A (Park *et al.*, 2015), cataracts (Wu *et al.*, 2013), hereditary tyrosinaemia type I (Yin *et al.*, 2014) and Duchenne muscular dystrophy (Long *et al.*, 2014). Moreover, the system is actively used for the removal of viruses such as hepatitis B (Ramanan *et al.*, 2015) and HIV (Ebina *et al.*, 2013) in human cell line. Beyond genome editing the technology is successfully applied for genome regulation (Fineran and Dy, 2014), imaging (Chen *et al.*, 2013), epigenetic regulation (Thakore *et al.*, 2015), fusion protein delivery (Mali *et al.*, 2013) (Fig. 6.1). However, to target any region in large genome, the binding of Cas9 to secondary genomic sites been remained a major complication in this field.

A number of novel strategies to measure and enhance genome wide specificity of Cas9 have been described in last three years which are comprehensively summarized in recent reviews (Jamal *et al.*, 2016).

In this chapter we discuss Cas9 specificity profiled by various biased and unbiased methods, and their advantages and disadvantages. We focus on molecular mechanisms of Cas9 targeting, binding and cleavage that defines Cas9 specificity, Cas9 and guide sequence engineering for improving Cas9 specificity (Fig. 6.2). The optimization and consideration of these approaches will greatly benefit the genome editing field to be broadly applied in therapies.

CRISPR off-targets

RNA guided endonuclease (RGENs) are now regarded as important member of engineered nucleases. Certain applications, e.g. therapeutic genome editing in human stem cells, generation of homozygous cell lines for evaluating fundamental genetic variations, demands high level of precision and specificity (Soldner *et al.*, 2011). These

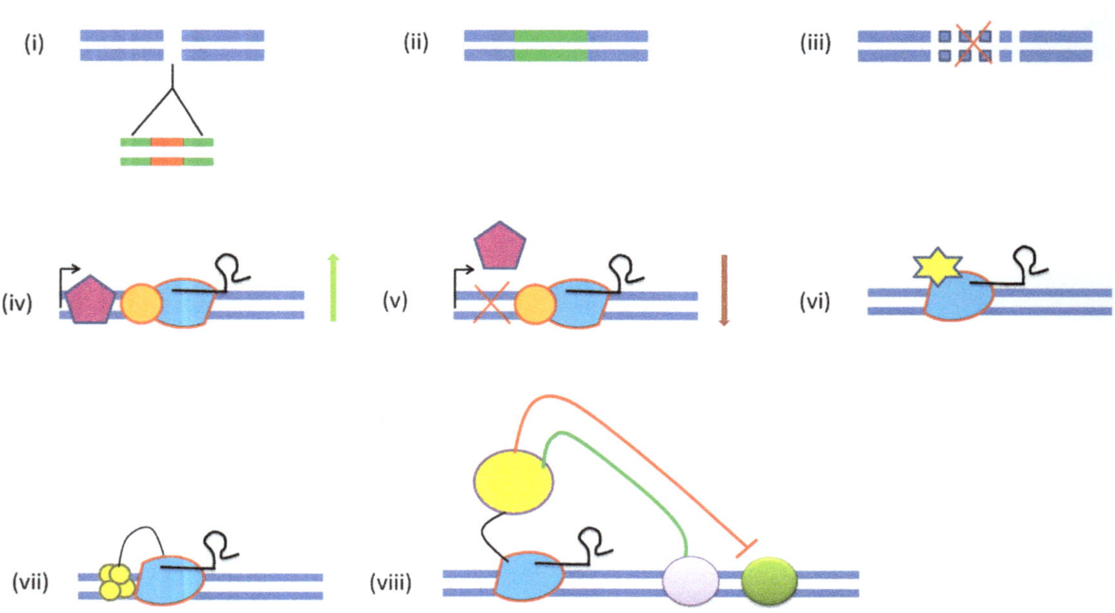

Figure 6.1 Application of CRISPR/Cas9/dCas9 based (A) gene disruption of wild type locus by using HDR template with desired mutation. (B) Gene knock-in to the target site by HR using HDR. (C) Gene knock-out caused by erroneous NHEJ pathways to rectify the DSB induced by Cas9. (D) dCas9 based up-regulation or transcriptional activation of target gene using dCas9 fused with domains like VP64 or omega. (E) Gene silencing or down-regulation of gene using dCas9 fused with repressor like KRAB. (F) Imaging of target genomic site using a colouring compound such as a fluorophore. (G) Antibody epitope tagging or any other fusion protein delivery. (H) Epigenetic modification (histone demethylation repress gene expression or histone acetylation activate gene expression).

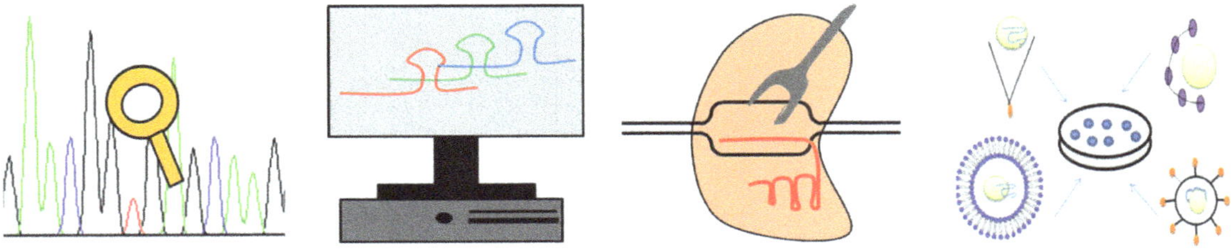

Figure 6.2 Schematic representation of the strategies used to increase CRISPR/Cas9 specificities, (A) analysis of off-target sites using biased and unbiased methods. (B) Web based designed of gRNA with minimum off-targets in the genome. (C) Engineering of Cas9 and gRNA. (D) Delivery methods of Cas9–sgRNA complex to allow transient expression of Cas9-sgRNA ribonucleoprotein complex.

nucleases with off-target DSB activity could induce undesirable mutations with potentially deleterious effects, an unacceptable outcome in most clinical settings. Similarly, due to the permanent nature of genomic modifications, specificity is of paramount concern to sensitive applications such as for studies aimed at linking specific genetic variants with biological processes, disease phenotypes, clinical applications and gene therapy. An ideal nuclease engineered nuclease should have single site cleavage specificity, but Cas9 has been shown to possess off-target cleavage activities.

Analysing off-target effects

Prediction of off-target sites is difficult, as the off-target cleavage is not fully understood. Various methods including restriction fragment length polymorphism (RFLP), PCR amplification of off-target sites followed by next generation sequencing (NGS), or mismatch sensitive nuclease (Surveyor assay and T7 endonuclease assay). The first method used for detection of off-targets was RFLP in cells (Urnov *et al.*, 2005). Surveyor assay and T7 endonuclease assays have been used to detect indel mutations generated as a result of erroneous pathway NHEJ in a population of cells.

Sanger sequencing is used to reveal mutations at off- and on-target sites induced by nucleases. Similarly, high-throughput sequencing predicts indel mutation (frequency ranging from 0.01% to 1%) up to hundreds at off- and on-target sites. However, a great deal of precautions should be taken to remove the false positive clones obtained during PCR artefacts (Cho *et al.*, 2014). These endonucleases recognize and cleave non-perfectly matched DNA (heteroduplexes) that are obtained as a result of annealing of wild-type and mutant DNA sequences. These digested DNA are then run on gel. The size and intensity of the bands can provide a clue about the mutation frequency. The T7E1 endonuclease assay is more sensitive than CEL-I to indels with deletion substrates (Vouillot *et al.*, 2015), while Surveyor is suited for detecting single nucleotide changes.

Genome wide assays for interrogation of off-target sites

The methods used to detect off-target sites are mostly based on binding and cleavage of these nucleases either in cell or out of cells (*in vitro*) are described in Fig. 6.3.

In-cell

The sequencing techniques such as sequencing the entire genome of the edited cells, known as whole genome sequencing, can be used to profile Cas9 specificity (Iyer *et al.*, 2015; Smith *et al.*, 2014). The sequencing result of the genome pre- and post-editing can reveal about the generation of novel mutation near the target site. This platform is useful when applied to single cell, clones non-mosaic F1 genome edited animals. However, the method is limited owing to low sensitivity; especially off-target with low frequencies are not detected, but also off-target in large population of cells (Smith *et al.*, 2014; Veres *et al.*, 2014).

The method is unbiased to gauge specificity of engineered nucleases for single cell-clones. However, the poor sensitivity to off-target sites especially those which occur at low frequency in a

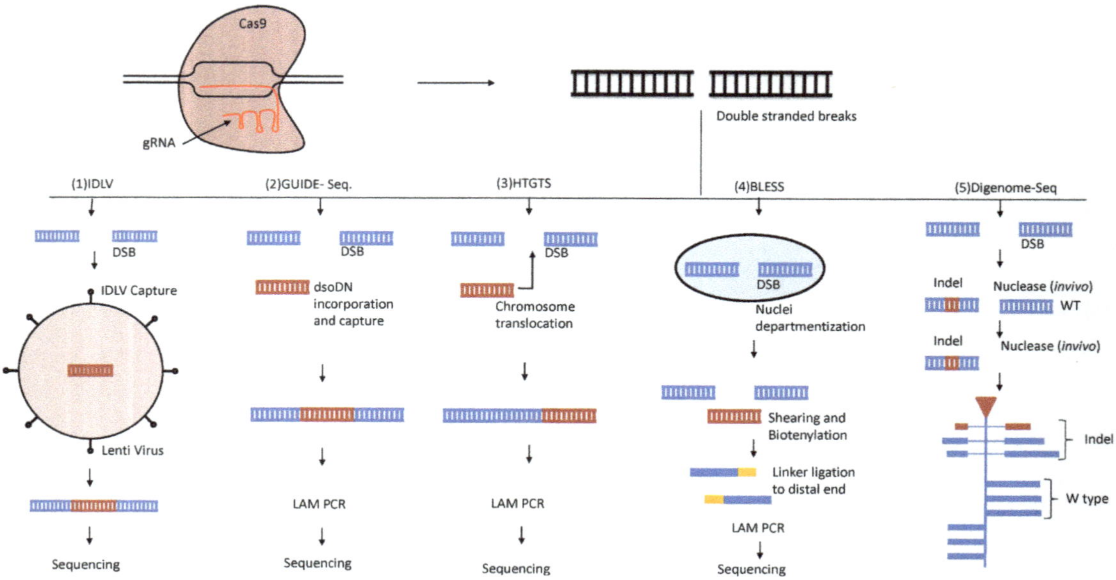

Figure 6.3 Schematic overview of off-target detection methods, the mechanism of each method is described in the text.

cell population (Tsai and Joung, 2014). DSB capturing method is based on incorporation of IDLV into the DSBs via non-homologous end joining (NHEJ) and tags those transient DSBs. The tagged sites are recovered by linear amplification-mediated PCR (LAM-PCR) and then mapped using high-throughput sequencing. The advantage of the IDLV capture method is that it can recognize the DSB induced by nuclease. However, the technique has some limitations such as less integration efficiencies requiring positive selection to resolve, and also random integration into sites free of nuclease induced DSBs (Gabriel *et al.*, 2011).

BLESS

BLESS (breaks labelling, enrichment on streptavidin and next-generation sequencing) is based on *in vivo* DNA cleavage, but *in vitro* DSB capturing detection method that involves fixation of cells, chromatin purification, nuclease digestion, ligation with biotinylated linkers and targeted deep sequencing. The process is beneficial as no exogenous bait is introduced to the cells also it can detect DSBs from tissues in *in vivo* (Ran *et al.*, 2015). The process is limited as the DSB captured by BLESS is at one and specific time at a specific time, during fixation but cannot detect the DSBs formed pre-fixation. The complex and time consuming protocol and also the requires a large number of cells (more than 10 million cells) are the disadvantages associated with this technique.

GUIDE-seq

GUIDE-seq (genome-wide unbiased identification of double stranded breaks enabled by sequencing) is a method used to detect the off-target effects in cells (Tsai *et al.*, 2015). The method involves the incorporation of a blunt, end-protected double-stranded oligodeoxynucleotide (dsODN) tag, specific amplification of the tag followed by high-throughput sequencing. The sequencing reads are compared with reference genome to locate the DSBs. The method is simple, efficient and sensitive, and can detect off-target sites that are mutagenized with frequencies of 0.1% or lower in a population of cells. The method precisely detects off-target sites that were not detected by other tools (Heigwer *et al.*, 2014) and also the computational pipelines available online for bioinformatics analysis (Tsai *et al.*, 2016). However, the disadvantage of the method is the requirement of cost-effective delivery method of dsODN to avoid its toxicity in some cell lines as observed.

HTGTS

HTGTS (high-throughput genome-wide translocation sequencing) (Frock *et al.*, 2015; Hu *et al.*, 2015) is an unbiased method for off-target site detection using known site of the genome as bait to capture prey DNA sequences that are translocated to on-target sites in live cells. The genome wide libraries containing bait–prey interactions are generated and are cloned using LAM-PCR, adapter ligation and PCR amplification of library followed by deep sequencing.

The drawback of this method is that translocation events occur very rarely depending on the three dimensional organization of genome, since translocations occur frequently on the same chromosome or a chromosome that lies in close proximity.

In vitro genome-wide assay for the detection of off-targets sites

The genomic DNA free from cells can be used to analyse the off-target effects *in vitro*.

Digenome-seq

The method used known as digested genome sequencing (Digenome-seq) is based on cleaving genomic DNA *in vitro* with Cas9 resulting in the same 5′ end at cleavage sites, adapter ligation followed by WGS. The cleaved sites are identified by read alignment (Kim *et al.*, 2015). The method offers advantage over GUIDE-seq as it can be applied to any sort of cell type without considering factors such as epigenetics, chromatin or subnuclear localization because digestion is performed on genomic DNA *in vitro*, and is much more sensitive than GUIDE-seq because increasing the concentration of Cas9–gRNA complex will enable the detection of sites cleaved with low frequency. The problem associated with Digenome-seq is the sequencing cost, moreover the method still needs to determine the difference between spontaneous mutation and that of the mutation induced by the off-target effect of the nuclease, so to this end the sequencing reads must be compared with other method to validate the cleaved sites.

SELEX

SELEX (systematic evolution of ligands by exponential enrichment) is an *in vitro* method based on the investigation of Cas9 binding to the target site rather than target cleavage. A set of oligonucleotides matching target sites in the genome are used and the bound sequences are then searched to a library of oligonucleotides to anticipate potential off-target sites under controlled conditions (Gabriel *et al.*, 2011).

ChIP-seq

ChIP-seq (chromatin immunoprecipitation) uses the binding of dCas9 to target site to asses off-target sites (Frock *et al.*, 2015; Wu *et al.*, 2014). Although ChIP is a useful unbiased method, unfortunately one should not correlate this method for Cas9 and dCas9 as DNA binding and cleavage are independent events, with cleavage more stringent than binding.

In summary, each and every method has disadvantages so using these methods in complementary manner will help understand potential off-target sites and its effects that will in turn help to develop novel approaches with optimized specificity. The biased method depends on bioinformatics algorithms to identify off-target sites, and on the other hand the unbiased method captures DSB generated by nucleases. The biased method is complemented by the unbiased method as it fails to predict about large number of off-target sites. For example, using *in vitro* test, off-target sites can be discovered (e.g. Digenome-seq), in cells (e.g. HTGTS, GUIDE-seq), or *in vivo* (e.g. BLESS) prior to validation. The combinations of these methods HTGTS, GUIDE-seq, and Digenome-seq revealed a set of potential off-target sites used for one sgRNA targeting VEGF-A site. In this regard the use of larger datasets can accurately define off-targeting *in vivo* and in cell that in turn will help in the development of bioinformatics tools used to design and select target sites.

gRNA designing tools

Experimental identification of off-target sites by *in vitro* or *in vivo* methods is laborious. Therefore, *in silico* prediction tools are broadly applied. The first generation *in silico* tools were mostly designed according to the degree of similarity between the target site and secondary sites in the genome. The recently developed tools predict unique genomic sites with minimum sequence homology to the secondary genomic sites (Frock *et al.*, 2015; Heigwer *et al.*, 2014).

While designing an sgRNA the search for potential off-targets in the genome is important.

The web based tools developed by research groups are described in Table 6.1. The basic feature of each tool is to look for PAM and seed sequence in a genomic DNA. Tools such as CasOT (available online: http://eendb.zfgenetics.org/casot/) and sgRNAcas9 (available online: www.biootools.com) also predict the number and position of mismatches.

Cas-OFFinder [available online: www.rgenome.net/cas-offinder/ (accessed on 14 October 2015)] and COSMID [available online: https://crispr.bme.gatech.edu (accessed on 14 October 2015)] incorporate input sequence and search for 20 nt target sequences near to the canonical or non-canonical PAM (NGG or NRG), giving output off-target sites.

The web tools search off-target sites based on mismatches not on the basis of indels. However, it has been showed that gRNA with few indels may induce cleavage (CRISPR/Cas9 systems have off-target activity with insertions or deletions between target DNA and guide RNA sequences). Therefore, to look for off-target sites in a given sequence it is also necessary to check presence of indels as well as nucleotide mismatches, tools like Cas-OFFinder, CRISPRdirect and COSMID provide these features. The online tools can be applied for limited organisms but the off-line tools selects target sites from many organisms.

How to improve the specificity of CRISPR/Cas9 system?

Off-target mutations can be caused both by Cas9 and gRNA. Therefore, improvement of the current generation of CRISPR/Cas platform is mostly focused on these two factors. Several efforts aiming to improve CRISPR specificity should be performed prior to using CRISPR/Cas9-mediated genome editing, including the rational design of gRNA, selection of eligible CRISPR nuclease, choice of suitable target sites and delivery of Cas9-sgRNA into cells followed by rapid degradation

Table 6.1 Web based tools to design gRNA

Name of tool	No. of organism genomes supported	Length of input sequence	Website
ZiFiT	9	Target sequence (more than 1 kb)	http://zifit.partners.org/ZiFiT/Disclaimer.aspx
Cas-OFFinder	More than 30	Designed guide sequence (15 to 25 nt)	http://www.rgenome.net/cas-offinder/
E-CRISP	More than 30	Target sequence (more than 10 kb)	http://www.e-crisp.org/E-CRISP/designcrispr.html
CHOPCHOP	More than 25	More than 10 kb	https://chopchop.rc.fas.harvard.edu
CasOT	Any	More than 10 kb	http://eendb.zfgenetics.org/casot/
Blenching	22		https://benchling.com
CRISPR-P	More than 25 plants	More than 5 kb	http://cbi.hzau.edu.cn/crispr/
COSMID	More than 8	Out sgRNA designed is 10 to 55 nt	https://crispr.bme.gatech.edu/
DNA 2.0 CRISPR gRNA Design Tool	Any	Input target sequence more than 10 kb	https://www.dna20.com/eCommerce/cas9/input
CRISPR-direct	More than 20	Input target sequence more than 10 kb	http://crispr.dbcls.jp/
MIT CRISPR design tool	More than 10	Input target sequence more than 10 kb	http://crispr.mit.edu/
CRISPR-ERA	More than 10	Input target sequence more than 5 kb	http://crispr-era.stanford.edu/

of the nuclease. Additionally, understanding the molecular mechanism of binding and cleavage of Cas9–gRNA complex can improve the specificity of nuclease.

Specificity achieved by using modified gRNA

The truncated gRNAs (tru-gRNAs), with shorter sequences (<20 nucleotides, usually 17 or 18) can effectively improve the specificity of Cas9 (Fu *et al.*, 2014b). The off-target mutagenesis was largely reduced about 5000-fold by using tru-gRNA (Fu *et al.*, 2014b). GUIDE-seq showed that numbers of off-target cleavage sites across the genome were decreased by 2- to 5-fold for the same sites digested with Cas9 guided by full length gRNA. Intriguingly, truncation at the 5′ end of gRNA increase the on-target specificity, while truncation at the 3′ end of gRNA reduces the on-target cleavage activity of Cas9 nucleases (Fu *et al.*, 2013; Hwang *et al.*, 2013). Moreover, truncation of the standard gRNA by one to three nucleotides showed comparable activity to that of the full length gRNA however deleting more than three nucleotides (16 nucleotides gRNA) resulted in the loss of activity (Fu *et al.*, 2014b).

Recently tru-RFNs (combination of tru–gRNA pair with RNA guided FOKI nucleases) showed high activity in human cancer cell lines and embryonic stem cells with reduced undesirable off-target effects (Wyvekens *et al.*, 2015). The tru-gRNA provides Cas9 with the necessary energy for binding to target site and reduces the affinity to off-target site. Usually, the full length gRNA provides more energy than the required, allowing the Cas9 to bind to unspecific sequences. In addition, the efficiency of tru-gRNA can be further enhanced when truncated gRNAs are conjugated with paired Cas9 nickases (Fu *et al.*, 2014b).

Another strategy is using the gRNA containing additional guanine bases at the 5′end (5′-GGX20 or 5′-GGGX19) makes it more specific to the target by an unknown exact mechanism. the possible mechanism might be changing the gRNA stability and secondary structure or stabilizing protein interaction with 5′ end of the gRNA (Cho *et al.*, 2014; Kim *et al.*, 2015) (shown in Fig. 6.1C). In some cases it has been reported that these extensive gRNA exhibit lower on-target activity compared to gRNA of normal length (Cho *et al.*, 2014; Kim *et al.*, 2015). In another strategy the introduction of an A–U base pair to the gRNA scaffold stabilizes the gRNA scaffold. Besides, extension of the base pair in the hairpin increases the binding rate of gRNA to target site and finally enhances its on-target specificity (Chen *et al.*, 2013).

The synthesis of gRNA with chemical modification at 5′ and 3′ termini increased Cas9 specificity allowing successful genome editing in human T cells and CD34+ haematopoietic stem and in progenitor cells the modification consists of adding 2′-O-methyl (M), 2′-O-methyl 3′phosphorothioate (MS), or 2′-O-methyl 3′ thioPACE (MSP) to the last three nucleotides at both 5′ and 3′ ends (Hendel *et al.*, 2015). However using these modified gRNA revealed off-target activity for four predicted sites using deep sequencing.

In another approach Rahdar *et al.* (2015) synthesized a 29-nucleotide synthetic crRNA (ScrRNA) with methyl modifying the 5′ end base pairs and bases and interspersed 2′-fluoro modifications towards the 3′ end (Rahdar *et al.*, 2015). These modified guides although showed enhanced stability and specificity but showed a reduced the on-target effects.

Lesson learnt from biochemical, structural and functional studies

The dilemma of understanding the mechanism of how Cas9 recognize and cleave the target DNA is of pivotal importance to achieve enhanced specificity in genome editing. Recent good models describe a comprehensive mechanism and insights into the conformational changes when Cas9 recognize and binds to its target in complex with gRNA (Jiang *et al.*, 2015) and associated with gRNA and the target strand without (Nishimasu *et al.*, 2014) or with the PAM sequence (Anders *et al.*, 2014). Cas9 comprises of two lobes; a nuclease lobe containing two nuclease domains, namely HNH and Ruvc that performs the duty of double-stranded DNA cleavage, and a recognition lobe that binds sgRNA to the target DNA where HNH and Ruvc cleaves the complementary strand of DNA and the non-target DNA respectively (Nishimasu *et al.*, 2014). The nuclease lobe also interacts with protospacer adjacent motif (PAM) by mean of presence of a carboxyl terminal. A DNA sequence of 2–5 bp (mostly 5′-NGG-3′, where N can be any nucleotide) is crucial for CRISPR-based cutting of DNA located adjacent to the target sequence (protospacer). In the context of recognition the Cas9 interacts with repeat anti-repeat region of sgRNA when recognizing PAM proximal position on target DNA, and then recognizing the complementary sequence on the target DNA by means of 20 bp gRNA sequence. The protospacer interacting (PI) domain identifies the PAM sequence leading to the formation of R-loop. The formation of R-loop crucial for making gRNA-DNA-Cas9 (ternary complex) activates the NHN and RuvC domains to cleave the phosphate group on non-complementary and on complementary strands respectively (Jinek *et al.*, 2012).

PAM interactions are important for high-affinity Cas9 binding to target DNA (Jinek *et al.*, 2012). Cas9 in complex with gRNA is pre-organized to make PAM-interacting contacts (Jiang et *al.*, 2015). The interaction of Cas9 with PAM sequence on the target non-target strand changes the conformation resulting in the unwinding of DNA (Anders *et al.*, 2014). Moreover, future engineering of the PI domain may allow us to program the PAM specificity, improve the target site recognition fidelity, and increase the versatility of the Cas9 genome engineering platform.

Specificity achieved by Cas9 engineering

Cas9 mutant and its variant

The mutated Cas9 (Cas9n) with one point mutation, either H840A in the HNH domain or D10A in the RuvC domain, was found to be more specific and efficient for genome targeting purposes (Ran *et al.*, 2013) (shown in Fig. 6.1F). Cas9n has the ability to create offset nicks instead of a DSB at the target site guided by pair of gRNAs resulting in the formation of indel mutations (Cong *et al.*, 2013; Ran *et al.*, 2013). Additionally, a pair of double nicks at two sites by four customized gRNAs–Cas complex successfully deleted genomic fragments of up to 6 kb in HEK 293FT cells (Ran *et al.*, 2013). Cas9n reduces the off-targeting by 50- to 1000-fold without sacrificing the on-target effect (Ran *et al.*, 2013). Thus combination of two gRNAs is preferred when the specificity is of paramount importance (e.g. the creation of engineered cell lines). The delivery of two gRNAs reduces the likelihood that their off-target sites are close enough to cause double stranded breaks. This approach has led the foundation to edit stem cells for therapeutic purposes (Ran *et al.*, 2013).

It has been shown that Cas9n mutagenizes certain target sites at high frequency, presumably because the nicks created at these sites can be converted into double-strand breaks. While targeting by Cas9n the off-target mutation induced by the second gRNA in a pair is not precisely understand, to this end the off-target effects of both gRNA should be considered. The method discussed to detect the nuclease profile of Cas9n targeting RAG1 (recombination activating gene 1), HTGTS revealed the detection of few translocation junctions detected genome wide. The problem associated with the use of Cas9n is its high frequency of inducing point mutations at certain target sites (Cho *et al.*, 2014; Fu *et al.*, 2014b; Tsai *et al.*, 2014) and also low efficiency of targeting with some gRNAs.

Specificity achieved by alanine substitution in Cas9

Recently researcher of Zhang group developed eSpCas9 and eSaCas9 (enhanced specificity SpCas9 and SaCas9) respectively based on structural engineering of SpCas9 and SaCas9 (Slaymaker *et al.*, 2016). These nucleases are different from their parent nuclease by three and four codon substitutions respectively with increased specificity detected by BLESS. These mutations are presumably weakening the interaction and binding of Cas9 proteins to the non-target DNA.

An analogous approach used by researchers developed another variant of Cas9 described as HF (High Fidelity)-SpCas9 based on previous structural profile of Cas9 (Kleinstiver *et al.*, 2015). This type of nuclease contains four alanine substitutions in the residues that interact with phosphate backbone of the target DNA. The HF-SpCas9 possesses high efficiency at on-target compared with Cas9 when observed for gRNA targeting sites where the wild type Cas9 showed off-target mutation. The disadvantage of these variants is the induction of mutation at some off-target sites.

Another strategy involved the engineering of Cas9 DNA-binding ability by fusing it with DNA-binding domain to produce a chimera complex known as Cas9-pDBD. This complex reduced improved the precision and specificity nearly 150-fold, this is because binding of the domain increases the concentration of SpCas around the target site (Bolukbasi *et al.*, 2015). The wide range of targeting is probably provided by skipping the formation of R-loop followed by PAM recognition (Szczelkun *et al.*, 2014).

Another approach is modifying the PAM recognition sequence in Cas9. PAM plays a key role in the recognition of target DNA sequence. This will help to narrow down target site selection and also confer specificity to this type of variant. This may be achieved by altering the protospacer interacting domain (PID) to recognize a different PAM sequence as described by (Fonfara *et al.*, 2014; Nishimasu *et al.*, 2014) where the Cas9 sequences from *Staphylococcus pyogenes* and from *Streptococcus thermophilus* CRISPR were 60% identical. These variants can be guided by dual gRNAs without affecting the specificity and efficiency (Fonfara *et al.*, 2014).

dCas9 specificity

The dead Cas9 generated as a result of double mutations, fused with effectors domains, is used for transcriptional modulation, chromosomal labelling and epigenetics regulation (Konermann *et al.*, 2013). The dCas9 lacks cleavage activity but retains the binding ability.

dCas9–FOKI

The fusion of FOKI domain to the carboxy terminal of zinc-finger nucleases (ZFNs) and transcription activator-like effector nucleases (TALENs) achieved mark specificity (Bogdanove and Voytas, 2011). Here wild type nuclease FokI domain fused to the amino terminal of catalytically inactive Cas9 (dCas9) protein known as RFNs. The dCas9 is itself unable to cut DNA, but the fused Fok1 endonuclease cuts the DNA as a dimer again assisted by two sgRNAs with closely juxtaposed target site. The process is proved to be more specific as the RFNs complex does not induce any DNA lesions. RFNs offer a better recognition fidelity and specificity of ≥ 140-fold greater than Cas9 and 4-fold lower to that of Cas9n (Guilinger *et al.*, 2014b). RFNs are governed by the half-sites bound by each gRNA/FokI–dCas9 complex in a particular relative orientation (PAMs-out) in which PAM site lies outside of the intervening spacer (14 to 17bp). The requirement of two gRNAs for precise orientation of the target sites and for correct spacing between half-sites is more inflexible for the dCas9–Fok1 fusion construct than that for Cas9 nickase (Tsai *et al.*, 2014). Such a tough requirement selection gRNAs will likely reduce the number of available on-target sites. Moreover, the dimerization of paired dCas9 binding to two precise genomic targetable half-sites should render dCas9–Fok1 fusion nucleases substantially improved on-target specificity.

Coupling tru-gRNA with dCas9-FOKI (tru-RFNS) reduced off-target mutagenesis by 40% compared to RFNs in human cells (Wyvekens *et al.*, 2015). These tru-RFNs reduce the possibility of point mutations, tru-gRNAs provide necessary energy for binding to target site compare to full length gRNA which provide extra energy, thus allowing the Cas9 to bind to unspecific sequences.

The non-target DNA (non-complementary to gRNA) released first after Cas9 cleavage, the use of single stranded donor DNA complementary to the non-target strand showed an increased rate of HDR compared to the donor complementary to the target strand (Richardson *et al.*, 2016).

RFNs strictly depend upon appropriately targeted gRNAs. In a broader sense, the development of dimeric Fok1–dCas9 fusions may prove a significant turning point towards practical mammalian genome manipulations with high efficiency and accuracy. However, a downside of both systems is that both double-nicking and dimeric Fok1-dCas9 is reported to have lower efficiency than wild type Cas9, and also the large size of these genes (more than 8 kb), which is hard to carry for the vectors (capacity of AAV is 4.2 kb). Moreover, the requirement for PAM sites within a specified region limits the possible target sites. To this point the improvement in RFN by engineering the FOKI domain like heterodomain FOKI and gRNA sequence can aid to reduce or eliminate the off-target effects observed in single gRNA guiding.

Looking for Cas9 orthologues and CRISPR proteins across the bacterial kingdom

Recently discovered Cas9 orthologue from *Staphylococcus aureus* (SaCas9) four times smaller than SpCas9 allow easily packaging into AAV for delivery (Friedland *et al.*, 2015; Ran *et al.*, 2015). The requirement of a guide sequence and PAM is different for SpCas9 and SaCas9. The former needs a guide sequence of 20 nt and 5′-NGG PAM, the later needs a guide sequence of 21 to 24 nt and 5′ NNGRRT as PAM sequence. SaCas9 guided by 5′ (NGG)RRT has been showed to induce lower indel mutations at off-target sites (Ran *et al.*, 2015).

The Cas9 discovered from *Neisseria meningitidis* (NmCas9) with a more complex PAM sequence (5′-NNNNGATT) and a 21 to 24 bp guide sequence has been used in mammalian cell genome editing (Mali *et al.*, 2013a; Müller *et al.*, 2016). The long PAM sequence limits the number of off-target sites, as the frequency of short PAM 5′-NGG occurs after every 8 bp whereas lengthy PAM 5;-NNNNGATT occur after 128 bp. However, the longer PAM sequence limits the target site range (Lee *et al.*, 2016).

The presence of CRISPR system in diverse prokaryotic kingdom allows scientists to discover CRISPR system across different bacterial genome. Recently discovered nuclease known as Cpf1 (CRISPR from Prevotella and Francisella 1) belongs to class 2 type V-A CRISPR system has been efficiently used for human genome editing (Zetsche *et al.*, 2015). The PAM sequence for Cpf1 is 5′TTN present upstream of the target site. The target DNA is cleaved at different position, producing a staggered cut unlike Cas9, which produces blunt-ended DNA (Fonfara *et al.*, 2014; Yamano *et al.*, 2016). The genome editing specificity of Cpf1 was found to be comparable to that of Cas9 profiled Digenome-seq and GUIDE-seq (Kim *et al.*, 2016).

C2c2 is a member of class2, type VI present in *Leptotrichia shahii* provides immunity against RNA viruses. Researchers from MIT have used it for programmable repression of target RNA guided by crRNA (Abudayyeh *et al.*, 2016). The C2c2 binds to target sites and acts as RNase inhibiting RNA. However, the system is not completely known and the factors that govern the specificity (PAM, protein structure, and guide sequence) are not completely elaborated.

Concentration of Cas9-gRNA

The rate of off-target mutagenesis also increases with increase in the concentration of TALEN as described by Guilinger *et al.* (2014a). This reflects that the lower concentration of Cas9 reduces the extra energy for DNA binding to avoid the binding to off-target sites *in vitro* and in cells (Fu *et al.*, 2014a; Hsu *et al.*, 2013; Pattanayak *et al.*, 2013).

There are several methods for the delivery of Cas9-gRNA into the cells; the most common method is the transfection of plasmid containing gRNA and Cas9 cassettes. The method for delivery of the CRISPR system is of great concern as it allows controlled optimum concentration of Cas9 and sgRNA in a cell that in turn is important for increasing the specificity (Hsu *et al.*, 2013; Pattanayak *et al.*, 2013). Recently, a positive correlation was observed between Cas9/sgRNA expression level and mutagenesis efficiency in plants (Mikami *et al.*, 2015), as indicated by increased specificity when lowering the amount of plasmid DNA in transfection. While in excessive enzyme concentration, mismatches are tolerated (Pattanayak *et al.*, 2013). However, it is noteworthy that reducing the Cas9/gRNA complex also decreases the on-target site cleavage efficiency (Pattanayak *et al.*, 2013; Ran *et al.*, 2013). The delivery of purified Cas9 in complex with gRNA via electroporation or lipofection achieved efficient genome editing with low off-target effects (Cho *et al.*, 2013; Lin *et al.*, 2014; Zuris *et al.*, 2015).

The use of a weak or inducible promoter is thought to regulate the Cas9 expression to optimal level. The constitutive expression of these endonuclease increases the on-targeting editing but might also lead to higher off-target effects, as observed for zinc finger nucleases (ZFNs) (Gaj *et al.*, 2012). The expression timing of Cas9 is an additional factor affecting off-target, because short lived and timed delivery of Cas9 allowed minimum off-target effects (Lin *et al.*, 2014).

The regulation of Cas9 concentration at transcriptional and post-transcriptional level is important to allow successful genome editing; using low or inducible promoter can achieve this control expression. To achieve post-transcriptional control of Cas9 in cells Davis *et al.* (2015) developed Cas9 regulated by a small molecule known as intein that undergoes protein splicing when 4-hydroxyamoxefin is available. This conditionally active chimera achieved successful genome editing in cell line with 25-fold for 11 off-target sites already assessed (Davis *et al.*, 2015).

It has been shown that cell cycle synchronization of the nuclease in G2 increases HDR efficiency while reducing unwanted NHEJ events when supplied with RNP (Lin *et al.*, 2014). Researchers have used inhibitors of the NHEJ repair pathway. Maruyama *et al.* (2015) found that genome editing efficiency was increased up to 19-fold by preventing NHEJ when the DNA ligase IV (a principal enzyme involved in NHEJ repair pathway) was antagonized by its inhibitor Scr7.

What else should be considered in avoiding off-target mutagenesis?

Cell fate

The DSB induced Cas9 and its repair is highly cell specific depending on the cell DNA repair machinery (Duan *et al.*, 2014), for instance, the vulnerability of off-target mutation is high in transformed human cell lines where the repair system is not intact compared to healthy human cell line with a regulated DSB repair pathway (Smith *et al.*, 2014; Veres *et al.*, 2014). The epigenetic state of the cell also impacts the off-target effects; DNA methylation at CpG island may interfere with the binding of Cas9. The methods to detect off-target effects have been applied to cancer cell lines such as HEK293, U2OS, and K562 where the nuclease-induced off-targets were detectable compare to stem cells (ES, iPCS) where the off-target were not detectable. The higher number of off-targets in cancer cell lines might reflect the abnormal DNA repair mechanism, as observed by GUIDE-seq (enables genome-wide profiling of off-target cleavage by CRISPR/Cas nucleases) where DSB were observed in the absence of CRISPR/Cas9.

Protospacer adjacent motif (PAM)

The protospacer-adjacent motif (PAM), three nucleotides (NGG) canonical PAM sequence, identified by one of the Cas9 domains is strictly required to be immediately next to the 3′ end of the target sequence. The PAM binding is required for opening and cleavage of the target site (Nishimasu *et al.*, 2014). The number of possible potential gRNAs for a gene is proportional to the length of the gene. The *S. pyogenes* PAM site is repeated every 8 nucleotide on average. Using the frequency of 'GG' equal to 5.21% in the reference human genome there would be an expected 161,284,793 NGG PAM sites in the human genome, or roughly one 'GG' dinucleotide every 42 bases. The most important triplet nucleotide sequence (NGG) that itself is not included in the sgRNA sequence, is identified by one domain of Cas9 enzyme (Nishimasu *et al.*, 2014). The PAM sequence NRG (where R can be G or A) is identified by Cas9-mediated DNA cleavage at the human EMX locus (Hsu *et al.*, 2013; Jiang *et al.*, 2013) although with lower binding efficiency (one-fifth) than canonical PAM sequence.

Therefore, the designing of gRNA should be based on the fact that there should be minimum or no homology between off-target and on-target sites and the off-target should not be followed by the canonical or non-canonical PAM.

PAM distal versus PAM proximal nucleotides

The seed sequence (5–12 nt) near the PAM sequence is critical for Cas9-sgRNA binding; mismatches in the PAM-distal nucleotides are less important and generally have less effect on disturbing the sgRNA–target DNA hybrid than those in the PAM-proximal nucleotides (PAMPNs). Therefore, it suggests that PAM-proximal nucleotides are major determinants of Cas9 binding with target sites (Cong *et al.*, 2013; Fu *et al.*, 2013; Jinek *et al.*, 2012; Kuscu *et al.*, 2014; Mali *et al.*, 2013c). Moreover, the seed sequence with multiple U also increases the specificity as more likely resulting in decrease sgRNA abundance also terminating the transcription of sgRNA (Wang *et al.*, 2014; Wu *et al.*, 2014).

GC content of sgRNA

It has been shown that mutagenesis efficiency is associated with GC content of PAM proximal nucleotides (PAMPNs) in gRNA. gRNA with three or fewer GCs in PAMPNs rarely reach 60% heritable mutation rate compare to more than 60% heritable mutation of gRNA with four GCs in PAMPNs (Ren *et al.*, 2014). This is because high GC content stabilizes the sgRNA–DNA hybrid allowing maximum Cas9 efficiency. The sgRNA containing G at the first position of the seed sequence is preferable; on the other hand the presence of C but not guanine at fifth position proximal to PAM and A in the middle of sgRNA is ideal.

Future directions

The specificity is of paramount importance in therapeutic applications. The problems of poor specificity are overcome by the latest approaches discussed in this chapter. The progress to determine the methods to detect the off-target mutations is remarkable, but still there is room for improvement, especially improving the sensitivity for the detection of off-target sites, as these biased and unbiased methods have detection limit of 0.1% yet these method are insensitive to certain off-target sites. To this end, the development of more accurate and highly sensitive methods that can detect the off-target site below 0.1% is desperately needed. Combination of these methods using the same gRNA and same target site would reinforce the detection of off-target sites.

The engineering of Cas9 (discovery of eSpCas9 and SpCas9-HF) and gRNA has further strengthened the specificity. These new variants increased reduced tolerance to mismatched DNA sites. Moreover the discoveries new nucleases Cpf1, C2c2, Cas10 and NgAgo equipped the genome editing toolbox to be used broadly in genome engineering. However the off-target effects of these nucleases are not described yet. Other approaches like use of safe and efficient delivery methods to allow transient expression of Cas9 and sgRNA will reduce the safety risk during clinical and therapeutic application built on genome-editing nucleases. Comprehensive consideration and propagation of these approaches will greatly facilitate genome editing technology to achieve a marked specificity and efficiency.

Glossary

CRISPR: A system present in bacteria providing immunity against invading agents such as viruses and plasmids.

Cas9: CRISPR-associated protein component of CRISPR system responsible for breaking double-stranded DNA when guided by gRNA.

CrRNA: CRISPR RNA is the transcribed product of spacer–repeat–spacer. The spacer is complementary to the target DNA sequence.

tracrRNA: Stands for trans-activating CrRNA, is the product of DNA sequences present near the CRISPR locus. The activation of CRISPR system requires the hybridization of crRNA into tracrRNA to form guide RNA (gRNA).

Protospacer: The target DNA sequence where Cas9 produces cleavage.

PAM: PAM stands for protospacer adjacent motif and this is the sequence recognized by Cas9 for licensing the DNA binding and cleavage.

NHEJ: Non-homologous end joining is a repair pathway used by the cell to rectify the double-stranded break induced by nucleases in the absence of a homology-directed repair template. The resultant correction may bring point mutations into the genome, often leading to gene knock-out.

HDR: Homology directed repair is another pathway used when a donor DNA template is provided containing the gene of interest to knock-in into the target after the DSB induced by nuclease. This type of pathway results in the formation of knock-in/-out.

Cas9 nickases: A mutant version of Cas9 with a mutation in one of its DNA-cleaving domains, the resultant Cas9 nick single strand of double-stranded DNA.

Dead Cas9 (dCas9): Engineering Cas9 by mutation in both nuclease domains; as a result it cannot cut the DNA but retains the DN- binding ability.

High-throughput sequencing: A DNA sequencing platform with high accuracy compared to Sanger sequencing method, which can be applied to sequence with short reads.

IDLV: The incorporation of integrase-deficient lentivirus (IDLV) into DSB in live cells can capture the DSB sites in live cell. These DSB are tagged and PCR amplified and analysed using high-throughput sequencing.

ChIP-seq: Chromatin immunoprecipitation, a method profiling the genome-wide protein interaction with DNA, the DNA associated with protein is sheared into pieces, unlinked, purified and sequenced.

SELEX: Systematic evolution of ligands by exponential enrichment is a method that analyses the binding of Cas9 to the target site. A random pool of oligo matching target sites are used, the unbound set of oligos is removed and the bound oligo are PCR amplified and sequenced.

Acknowledgements

This work was supported by the National Natural Science Foundation of China (Grant No. 31371106,

31421064), Hubei Provincial Natural Science Foundation of China (Grant No. 36113011), Doctoral Fund of Ministry of Education of China (Grant No. 131012) and the Huazhong Agricultural University Scientific and Technological Self innovation Foundation (Program No. 52204-13002).

Conflict of interest

The authors disclose no conflict of interest.

References

Abudayyeh, O.O., Gootenberg, J.S., Konermann, S., Joung, J., Slaymaker, I.M., Cox, D.B., Shmakov, S., Makarova, K.S., Semenova, E., and Minakhin, L. (2016). C2c2 is a single-component programmable RNA-guided RNA-targeting CRISPR effector. *Science* aaf5573.

Anders, C., Niewoehner, O., Duerst, A., and Jinek, M. (2014). Structural basis of PAM-dependent target DNA recognition by the Cas9 endonuclease. Nature *513*, 569–573. http://dx.doi.org/10.1038/nature13579

Bhaya, D., Davison, M., and Barrangou, R. (2011). CRISPR-Cas systems in bacteria and archaea: versatile small RNAs for adaptive defense and regulation. Annu. Rev. Genet. *45*, 273–297. http://dx.doi.org/10.1146/annurev-genet-110410-132430

Bogdanove, A.J., and Voytas, D.F. (2011). TAL effectors: customizable proteins for DNA targeting. Science *333*, 1843–1846. http://dx.doi.org/10.1126/science.1204094

Bolukbasi, M.F., Gupta, A., Oikemus, S., Derr, A.G., Garber, M., Brodsky, M.H., Zhu, L.J., and Wolfe, S.A. (2015). DNA-binding-domain fusions enhance the targeting range and precision of Cas9. Nat. Methods *12*, 1150–1156. http://dx.doi.org/10.1038/nmeth.3624

Chang, N., Sun, C., Gao, L., Zhu, D., Xu, X., Zhu, X., Xiong, J.-W., and Xi, J.J. (2013). Genome editing with RNA-guided Cas9 nuclease in zebrafish embryos. Cell Res. *23*, 465–472. http://dx.doi.org/10.1038/cr.2013.45.

Chen, B., Gilbert, L.A., Cimini, B.A., Schnitzbauer, J., Zhang, W., Li, G.-W., Park, J., Blackburn, E.H., Weissman, J.S., and Qi, L.S. (2013). Dynamic imaging of genomic loci in living human cells by an optimized CRISPR/Cas system. Cell *155*, 1479–1491. http://dx.doi.org/10.1016/j.cell.2013.12.001.

Cho, S.W., Kim, S., Kim, Y., Kweon, J., Kim, H.S., Bae, S., and Kim, J.-S. (2014). Analysis of off-target effects of CRISPR/Cas-derived RNA-guided endonucleases and nickases. Genome Res. *24*, 132–141. http://dx.doi.org/10.1101/gr.162339.113

Cho, S.W., Lee, J., Carroll, D., Kim, J.-S., and Lee, J. (2013). Heritable gene knockout in Caenorhabditis elegans by direct injection of Cas9–sgRNA ribonucleoproteins. Genetics *195*, 1177–1180. http://dx.doi.org/10.1534/genetics.113.155853

Cong, L., Ran, F.A., Cox, D., Lin, S., Barretto, R., Habib, N., Hsu, P.D., Wu, X., Jiang, W., and Marraffini, L.A. (2013). Multiplex genome engineering using CRISPR/Cas systems. Science *339*, 819–823. http://dx.doi.org/10.1126/science.1231143

Davis, K.M., Pattanayak, V., Thompson, D.B., Zuris, J.A., and Liu, D.R. (2015). Small molecule-triggered Cas9 protein with improved genome-editing specificity. Nat. Chem. Biol. *11*, 316–318. http://dx.doi.org/10.1038/nchembio.1793

Deveau, H., Garneau, J.E., and Moineau, S. (2010). CRISPR/Cas system and its role in phage-bacteria interactions. Annu. Rev. Microbiol. *64*, 475–493. http://dx.doi.org/10.1146/annurev.micro.112408.134123

Duan, J., Lu, G., Xie, Z., Lou, M., Luo, J., Guo, L., and Zhang, Y. (2014). Genome-wide identification of CRISPR/Cas9 off-targets in human genome. Cell Res. *24*, 1009.

Ebina, H., Misawa, N., Kanemura, Y., and Koyanagi, Y. (2013). Harnessing the CRISPR/Cas9 system to disrupt latent HIV-1 provirus. Scientific Rep. *3*.

Fineran, P.C., and Dy, R.L. (2014). Gene regulation by engineered CRISPR-Cas systems. Curr. Opin. Microbiol. *18*, 83–89. http://dx.doi.org/10.1016/j.mib.2014.02.007

Fonfara, I., Le Rhun, A., Chylinski, K., Makarova, K.S., Lécrivain, A.-L., Bzdrenga, J., Koonin, E.V., and Charpentier, E. (2014). Phylogeny of Cas9 determines functional exchangeability of dual-RNA and Cas9 among orthologous type II CRISPR-Cas systems. Nucleic Acids Res. *42*, 2577–2590. http://dx.doi.org/10.1093/nar/gkt1074

Friedland, A.E., Baral, R., Singhal, P., Loveluck, K., Shen, S., Sanchez, M., Marco, E., Gotta, G.M., Maeder, M.L., and Kennedy, E.M. (2015). Characterization of *Staphylococcus aureus* Cas9: a smaller Cas9 for all-in-one adeno-associated virus delivery and paired nickase applications. Genome Biol. *16*, 1.

Friedland, A.E., Tzur, Y.B., Esvelt, K.M., Colaiácovo, M.P., Church, G.M., and Calarco, J.A. (2013). Heritable genome editing in C. elegans via a CRISPR-Cas9 system. Nat. Methods *10*, 741–743. http://dx.doi.org/10.1038/nmeth.2532

Frock, R.L., Hu, J., Meyers, R.M., Ho, Y.-J., Kii, E., and Alt, F.W. (2015). Genome-wide detection of DNA double-stranded breaks induced by engineered nucleases. Nat. Biotechnol. 33, 179–186. http://dx.doi.org/10.1038/nbt.3101

Fu, B.X.H., Hansen, L.L., Artiles, K.L., Nonet, M.L., and Fire, A.Z. (2014a). Landscape of target:guide homology effects on Cas9-mediated cleavage. Nucleic Acids Res. gku1102.

Fu, Y., Foden, J.A., Khayter, C., Maeder, M.L., Reyon, D., Joung, J.K., and Sander, J.D. (2013). High-frequency off-target mutagenesis induced by CRISPR-Cas nucleases in human cells. Nat. Biotechnol. *31*, 822–826. http://dx.doi.org/10.1038/nbt.2623

Fu, Y., Sander, J.D., Reyon, D., Cascio, V.M., and Joung, J.K. (2014b). Improving CRISPR-Cas nuclease specificity using truncated guide RNAs. Nat. Biotechnol. *32*, 279–284. http://dx.doi.org/10.1038/nbt.2808

Gabriel, R., Lombardo, A., Arens, A., Miller, J.C., Genovese, P., Kaeppel, C., Nowrouzi, A., Bartholomae, C.C., Wang, J., and Friedman, G. (2011). An unbiased genome-wide analysis of zinc-finger nuclease specificity. Nat. Biotechnol. *29*, 816–823. http://dx.doi.org/10.1038/nbt.1948

Gaj, T., Guo, J., Kato, Y., Sirk, S.J., and Barbas, C.F. (2012). Targeted gene knockout by direct delivery of zinc-finger nuclease proteins. Nat. Methods *9*, 805–807. http://dx.doi.org/10.1038/nmeth.2030

Guilinger, J.P., Pattanayak, V., Reyon, D., Tsai, S.Q., Sander, J.D., Joung, J.K., and Liu, D.R. (2014a). Broad specificity profiling of TALENs results in engineered nucleases with improved DNA-cleavage specificity. Nat. Methods *11*, 429.–435. http://dx.doi.org/10.1038/nmeth.2845

Guilinger, J.P., Thompson, D.B., and Liu, D.R. (2014b). Fusion of catalytically inactive Cas9 to FokI nuclease improves the specificity of genome modification. Nat. Biotechnol. *32*, 577.

Heigwer, F., Kerr, G., and Boutros, M. (2014). E-CRISP: fast CRISPR target site identification. Nat. Methods *11*, 122–123. http://dx.doi.org/10.1038/nmeth.2812

Hendel, A., Bak, R.O., Clark, J.T., Kennedy, A.B., Ryan, D.E., Roy, S., Steinfeld, I., Lunstad, B.D., Kaiser, R.J., and Wilkens, A.B., *et al.* (2015). Chemically modified guide RNAs enhance CRISPR-Cas genome editing in human primary cells. Nat. Biotechnol. *33*, 985–989. http://dx.doi.org/10.1038/nbt.3290

Hsu, P.D., Scott, D.A., Weinstein, J.A., Ran, F.A., Konermann, S., Agarwala, V., Li, Y., Fine, E.J., Wu, X., and Shalem, O., *et al.* (2013). DNA targeting specificity of RNA-guided Cas9 nucleases. Nat. Biotechnol. *31*, 827–832. http://dx.doi.org/10.1038/nbt.2647

Hu, J., Zhang, Y., Zhao, L., Frock, R.L., Du, Z., Meyers, R.M., Meng, F.-l., Schatz, D.G., and Alt, F.W. (2015). Chromosomal loop domains direct the recombination of antigen receptor genes. Cell *163*, 947–959. http://dx.doi.org/10.1016/j.cell.2015.10.016

Hwang, W.Y., Fu, Y., Reyon, D., Maeder, M.L., Kaini, P., Sander, J.D., Joung, J.K., Peterson, R.T., and Yeh, J.-R.J. (2013). Heritable and precise zebrafish genome editing using a CRISPR-Cas system. PloS One *8*, e68708. http://dx.doi.org/10.1371/journal.pone.0068708

Iyer, V., Shen, B., Zhang, W., Hodgkins, A., Keane, T., Huang, X., and Skarnes, W.C. (2015). Off-target mutations are rare in Cas9-modified mice. Nat. Methods *12*, 479. http://dx.doi.org/10.1038/nmeth.3408

Jamal, M., Khan, F.A., Da, L., Habib, Z., Dai, J., and Cao, G. (2016). Keeping CRISPR/Cas on-target. Curr. Issues. Mol. Biol. *20*, 1-20.

Jiang, F., Zhou, K., Ma, L., Gressel, S., and Doudna, J.A. (2015). A Cas9–guide RNA complex preorganized for target DNA recognition. Science *348*, 1477–1481.

Jiang, W., Bikard, D., Cox, D., Zhang, F., and Marraffini, L.A. (2013). RNA-guided editing of bacterial genomes using CRISPR-Cas systems. Nat. Biotechnol. *31*, 233–239. http://dx.doi.org/10.1038/nbt.2508

Jinek, M., Chylinski, K., Fonfara, I., Hauer, M., Doudna, J.A., and Charpentier, E. (2012). A programmable dual-RNA-guided DNA endonuclease in adaptive bacterial immunity. Science *337*, 816–821. http://dx.doi.org/10.1126/science.1225829

Khan, F.A., Pandupuspitasari, N.S., Chun-Jie, H., Ao, Z., Jamal, M., Zohaib, A., Khan, F.A., Hakim, M., and ShuJun, Z. (2016). CRISPR/Cas9 therapeutics: a cure for cancer and other genetic diseases. Oncotarget. *7*, 52541–52552.

Kim, D., Bae, S., Park, J., Kim, E., Kim, S., Yu, H.R., Hwang, J., Kim, J.-I., and Kim, J.-S. (2015). Digenome-seq: genome-wide profiling of CRISPR-Cas9 off-target effects in human cells. Nat. Methods *12*, 237–243.

Kim, D., Kim, J., Hur, J.K., Been, K.W., Yoon, S.-H., and Kim, J.-S. (2016). Genome-wide analysis reveals specificities of Cpf1 endonucleases in human cells. Nat. Biotechnol. *34*, 863–868.

Kleinstiver, B.P., Prew, M.S., Tsai, S.Q., Topkar, V.V., Nguyen, N.T., Zheng, Z., Gonzales, A.P., Li, Z., Peterson, R.T., and Yeh, J.-R.J. (2015). Engineered CRISPR-Cas9 nucleases with altered PAM specificities. Nature *523*, 481–485. http://dx.doi.org/10.1038/nature14592

Konermann, S., Brigham, M.D., Trevino, A., Hsu, P.D., Heidenreich, M., Cong, L., Platt, R.J., Scott, D.A., Church, G.M., and Zhang, F. (2013). Optical control of mammalian endogenous transcription and epigenetic states. Nature *500*, 472–476.

Koonin, E.V., and Makarova, K.S. (2009). CRISPR-Cas: an adaptive immunity system in prokaryotes. F1000. Biol. Rep. *1*.

Kuscu, C., Arslan, S., Singh, R., Thorpe, J., and Adli, M. (2014). Genome-wide analysis reveals characteristics of off-target sites bound by the Cas9 endonuclease. Nat. Biotechnol. *32*, 677–683.

Lee, C.M., Cradick, T.J., and Bao, G. (2016). The Neisseria meningitidis CRISPR-Cas9 system enables specific genome editing in mammalian cells. Mol. Ther. *24*, 645–654. http://dx.doi.org/10.1038/mt.2016.8

Li, D., Qiu, Z., Shao, Y., Chen, Y., Guan, Y., Liu, M., Li, Y., Gao, N., Wang, L., and Lu, X. (2013). Heritable gene targeting in the mouse and rat using a CRISPR-Cas system. Nat. Biotechnol. *31*, 681–683. http://dx.doi.org/10.1038/nbt.2661

Lin, S., Staahl, B.T., Alla, R.K., and Doudna, J.A. (2014). Enhanced homology-directed human genome engineering by controlled timing of CRISPR/Cas9 delivery. Elife *3*, e04766. http://dx.doi.org/10.7554/eLife.04766

Prevention of muscular dystrophy in mice by CRISPR/Cas9-mediated editing of germline DNA. Science *345*, 1184–1188. http://dx.doi.org/10.1126/science.1254445

Mali, P., Aach, J., Stranges, P.B., Esvelt, K.M., Moosburner, M., Kosuri, S., Yang, L., and Church, G.M. (2013a). CAS9 transcriptional activators for target specificity screening and paired nickases for cooperative genome engineering. Nat. Biotechnol. *31*, 833–838.

Mali, P., Esvelt, K.M., and Church, G.M. (2013b2013). Cas9 as a versatile tool for engineering biology. Nat. Methods *10*, 957–963. http://dx.doi.org/10.1038/nmeth.2649

Mali, P., Yang, L., Esvelt, K.M., Aach, J., Guell, M., DiCarlo, J.E., Norville, J.E., and Church, G.M. (2013c). RNA-guided human genome engineering via Cas9. Science *339*, 823–826. http://dx.doi.org/10.1126/science.1232033

Maruyama, T., Dougan, S.K., Truttmann, M.C., Bilate, A.M., Ingram, J.R., and Ploegh, H.L. (2015). Increasing the efficiency of precise genome editing with CRISPR-Cas9 by inhibition of nonhomologous end joining. Nat. Biotechnol. *33*, 538–542.

Mikami, M., Toki, S., and Endo, M. (2015). Parameters affecting frequency of CRISPR/Cas9 mediated targeted mutagenesis in rice. Plant. Cell. Rep., 1–9.

Müller, M., Lee, C.M., Gasiunas, G., Davis, T.H., Cradick, T.J., Siksnys, V., Bao, G., Cathomen, T., and Mussolino, C. (2016). Streptococcus thermophilus CRISPR-Cas9 systems enable specific editing of the human genome. Mol. Ther. *24*, 636–644.

Nishimasu, H., Ran, F.A., Hsu, P.D., Konermann, S., Shehata, S.I., Dohmae, N., Ishitani, R., Zhang, F., and Nureki, O. (2014). Crystal structure of Cas9 in complex with guide RNA and target DNA. Cell *156*, 935–949. http://dx.doi.org/10.1016/j.cell.2014.02.001

Niu, Y., Shen, B., Cui, Y., Chen, Y., Wang, J., Wang, L., Kang, Y., Zhao, X., Si, W., and Li, W. (2014). Generation of gene-modified cynomolgus monkey via Cas9/RNA-mediated gene targeting in one-cell embryos. Cell *156*, 836–843. http://dx.doi.org/10.1016/j.cell.2014.01.027

Park, C.-Y., Kim, D.H., Son, J.S., Sung, J.J., Lee, J., Bae, S., Kim, J.-H., Kim, D.-W., and Kim, J.-S. (2015). Functional correction of large factor VIII gene chromosomal inversions in hemophilia A patient-derived iPSCs using CRISPR-Cas9. Cell Stem Cell *17*, 213-220. http://dx.doi.org/10.1016/j.stem.2015.07.001

Pattanayak, V., Lin, S., Guilinger, J.P., Ma, E., Doudna, J.A., and Liu, D.R. (2013). High-throughput profiling of off-target DNA cleavage reveals RNA-programmed Cas9 nuclease specificity. Nat. Biotechnol. *31*, 839–843. http://dx.doi.org/10.1038/nbt.2673

Platt, R.J., Chen, S., Zhou, Y., Yim, M.J., Swiech, L., Kempton, H.R., Dahlman, J.E., Parnas, O., Eisenhaure, T.M., and Jovanovic, M. (2014). CRISPR-Cas9 knockin mice for genome editing and cancer modeling. Cell *159*, 440–455. http://dx.doi.org/10.1016/j.cell.2014.09.014

Rahdar, M., McMahon, M.A., Prakash, T.P., Swayze, E.E., Bennett, C.F., and Cleveland, D.W. (2015). Synthetic CRISPR RNA-Cas9--guided genome editing in human cells. Proc. Natl. Acad. Sci. U.S.A. *112*, E7110–E7117. http://dx.doi.org/10.1073/pnas.1520883112

Ramanan, V., Shlomai, A., Cox, D.B., Schwartz, R.E., Michailidis, E., Bhatta, A., Scott, D.A., Zhang, F., Rice, C.M., and Bhatia, S.N. (2015). CRISPR/Cas9 cleavage of viral DNA efficiently suppresses hepatitis B virus. Sci. Rep. *5*, 10833. http://dx.doi.org/10.1038/srep10833

Ran, F.A., Cong, L., Yan, W.X., Scott, D.A., Gootenberg, J.S., Kriz, A.J., Zetsche, B., Shalem, O., Wu, X., and Makarova, K.S. (2015). In vivo genome editing using Staphylococcus aureus Cas9. Nature *520*, 186–191. http://dx.doi.org/10.1038/nature14299

Ran, F.A., Hsu, P.D., Lin, C.-Y., Gootenberg, J.S., Konermann, S., Trevino, A.E., Scott, D.A., Inoue, A., Matoba, S., and Zhang, Y. (2013). Double nicking by RNA-guided CRISPR Cas9 for enhanced genome editing specificity. Cell *154*, 1380–1389. http://dx.doi.org/10.1016/j.cell.2013.08.021

Ren, X., Yang, Z., Xu, J., Sun, J., Mao, D., Hu, Y., Yang, S.-J., Qiao, H.-H., Wang, X., and Hu, Q. (2014). Enhanced specificity and efficiency of the CRISPR/Cas9 system with optimized sgRNA parameters in *Drosophila*. Cell. Rep. *9*, 1151–1162. http://dx.doi.org/10.1016/j.celrep.2014.09.044

Richardson, C.D., Ray, G.J., DeWitt, M.A., Curie, G.L., and Corn, J.E. (2016). Enhancing homology-directed genome editing by catalytically active and inactive CRISPR-Cas9 using asymmetric donor DNA. Nat. Biotechnol. *34*, 339–344.

Schwank, G., Koo, B.-K., Sasselli, V., Dekkers, J.F., Heo, I., Demircan, T., Sasaki, N., Boymans, S., Cuppen, E., and van der Ent, C.K. (2013). Functional repair of CFTR by CRISPR/Cas9 in intestinal stem cell organoids of cystic fibrosis patients. Cell Stem Cell. *13*, 653–658. http://dx.doi.org/10.1016/j.stem.2013.11.002

Slaymaker, I.M., Gao, L., Zetsche, B., Scott, D.A., Yan, W.X., and Zhang, F. (2016). Rationally engineered Cas9 nucleases with improved specificity. Science *351*, 84–88. http://dx.doi.org/10.1126/science.aad5227

Smith, C., Gore, A., Yan, W., Abalde-Atristain, L., Li, Z., He, C., Wang, Y., Brodsky, R.A., Zhang, K., and Cheng, L. (2014). Whole-genome sequencing analysis reveals high specificity of CRISPR/Cas9 and TALEN-based genome editing in human iPSCs. Cell. Stem. Cell. *15*, 12.

Soldner, F., Laganière, J., Cheng, A.W., Hockemeyer, D., Gao, Q., Alagappan, R., Khurana, V., Golbe, L.I., Myers, R.H., and Lindquist, S. (2011). Generation of isogenic pluripotent stem cells differing exclusively at two early onset Parkinson point mutations. Cell *146*, 318–331. http://dx.doi.org/10.1016/j.cell.2011.06.019

Szczelkun, M.D., Tikhomirova, M.S., Sinkunas, T., Gasiunas, G., Karvelis, T., Pschera, P., Siksnys, V., and Seidel, R. (2014). Direct observation of R-loop formation by single RNA-guided Cas9 and Cascade effector complexes. Proc. Natl. Acad. Sci. U.S.A. *111*, 9798–9803. http://dx.doi.org/10.1073/pnas.1402597111

Terns, M.P., and Terns, R.M. (2011). CRISPR-based adaptive immune systems. Curr. Opin. Microbiol. *14*, 321–327. http://dx.doi.org/10.1016/j.mib.2011.03.005

Thakore, P.I., D'Ippolito, A.M., Song, L., Safi, A., Shivakumar, N.K., Kabadi, A.M., Reddy, T.E., Crawford, G.E., and Gersbach, C.A. (2015). Highly specific epigenome editing by CRISPR-Cas9 repressors for silencing of distal regulatory elements. Nat. Methods *12*, 1143–1149. http://dx.doi.org/10.1038/nmeth.3630

Tsai, S.Q., and Joung, J.K. (2014). What's changed with genome editing? Cell. Stem. Cell. *15*, 3–4. http://dx.doi.org/10.1016/j.stem.2014.06.017

Tsai, S.Q., Topkar, V.V., Joung, J.K., and Aryee, M.J. (2016). Open-source guideseq software for analysis of GUIDE-seq data. Nat. Biotechnol. *34*, 483–483.

Tsai, S.Q., Wyvekens, N., Khayter, C., Foden, J.A., Thapar, V., Reyon, D., Goodwin, M.J., Aryee, M.J., and Joung, J.K. (2014). Dimeric CRISPR RNA-guided FokI nucleases for highly specific genome editing. Nat. Biotechnol. *32*, 569–576. http://dx.doi.org/10.1038/nbt.2908

Tsai, S.Q., Zheng, Z., Nguyen, N.T., Liebers, M., Topkar, V.V., Thapar, V., Wyvekens, N., Khayter, C., Iafrate, A.J., and Le, L.P. (2015). GUIDE-seq enables genome-wide profiling of off-target cleavage by CRISPR-Cas nucleases. Nat. Biotechnol. 33, 187–197. http://dx.doi.org/10.1038/nbt.3117

Tzur, Y.B., Friedland, A.E., Nadarajan, S., Church, G.M., Calarco, J.A., and Colaiácovo, M.P. (2013). Heritable custom genomic modifications in Caenorhabditis elegans via a CRISPR–Cas9 system. Genetics *195*, 1181-1185.

Urnov, F.D., Miller, J.C., Lee, Y.-L., Beausejour, C.M., Rock, J.M., Augustus, S., Jamieson, A.C., Porteus, M.H., Gregory, P.D., and Holmes, M.C. (2005). Highly efficient endogenous human gene correction using designed zinc-finger nucleases. Nature 435, 646–651.

Veres, A., Gosis, B.S., Ding, Q., Collins, R., Ragavendran, A., Brand, H., Erdin, S., Cowan, C.A., Talkowski, M.E., and Musunuru, K. (2014). Low incidence of off-target mutations in individual CRISPR-Cas9 and TALEN targeted human stem cell clones detected by whole-genome sequencing. Cell. Stem. Cell. *15*, 27–30. http://dx.doi.org/10.1016/j.stem.2014.04.020

Vouillot, L., Thélie, A., and Pollet, N. (2015). Comparison of T7E1 and surveyor mismatch cleavage assays to detect mutations triggered by engineered nucleases. G3: Genes Genomes Genet. *5*, 407–415.

Wang, T., Wei, J.J., Sabatini, D.M., and Lander, E.S. (2014). Genetic screens in human cells using the CRISPR-Cas9 system. Science *343*, 80–84. http://dx.doi.org/10.1126/science.1246981

Wiedenheft, B., Sternberg, S.H., and Doudna, J.A. (2012). RNA-guided genetic silencing systems in bacteria and archaea. Nature *482*, 331–338. http://dx.doi.org/10.1038/nature10886

Wu, X., Scott, D.A., Kriz, A.J., Chiu, A.C., Hsu, P.D., Dadon, D.B., Cheng, A.W., Trevino, A.E., Konermann, S., and Chen, S. (2014). Genome-wide binding of the CRISPR endonuclease Cas9 in mammalian cells. Nat. Biotechnol. *32*, 670–676. http://dx.doi.org/10.1038/nbt.2889

Wu, Y., Liang, D., Wang, Y., Bai, M., Tang, W., Bao, S., Yan, Z., Li, D., and Li, J. (2013). Correction of a genetic disease in mouse via use of CRISPR-Cas9. Cell. Stem. Cell. *13*, 659–662. http://dx.doi.org/10.1016/j.stem.2013.10.016

Wyvekens, N., Topkar, V.V., Khayter, C., Joung, J.K., and Tsai, S.Q. (2015). Dimeric CRISPR RNA-guided FokI-dCas9 nucleases directed by truncated gRNAs for highly specific genome editing. Hum. Gene Ther. *26*, 425–431. http://dx.doi.org/10.1089/hum.2015.084

Xie, F., Ye, L., Chang, J.C., Beyer, A.I., Wang, J., Muench, M.O., and Kan, Y.W. (2014). Seamless gene correction of β-thalassemia mutations in patient-specific iPSCs using CRISPR/Cas9 and piggyBac. Genome Res. *24*, 1526–1533. http://dx.doi.org/10.1101/gr.173427.114

Yamano, T., Nishimasu, H., Zetsche, B., Hirano, H., Slaymaker, I.M., Li, Y., Fedorova, I., Nakane, T., Makarova, K.S., and Koonin, E.V. (2016). Crystal structure of Cpf1 in complex with guide RNA and target DNA. Cell *165*, 949–962. http://dx.doi.org/10.1016/j.cell.2016.04.003

Yang, D., Xu, J., Zhu, T., Fan, J., Lai, L., Zhang, J., and Chen, Y.E. (2014). Effective gene targeting in rabbits using RNA-guided Cas9 nucleases. J. Mol. Cell. Biol. *6*, 97–99.

Yin, H., Xue, W., Chen, S., Bogorad, R.L., Benedetti, E., Grompe, M., Koteliansky, V., Sharp, P.A., Jacks, T., and Anderson, D.G. (2014). Corrigendum: Genome editing with Cas9 in adult mice corrects a disease mutation and phenotype. Nat. Biotechnol. *32*, 952–952.

Zetsche, B., Gootenberg, J.S., Abudayyeh, O.O., Slaymaker, I.M., Makarova, K.S., Essletzbichler, P., Volz, S.E., Joung, J., van der Oost, J., and Regev, A. (2015). Cpf1 is a single RNA-guided endonuclease of a class 2 CRISPR-Cas system. Cell *163*, 759–771. http://dx.doi.org/10.1016/j.cell.2015.09.038

Zuris, J.A., Thompson, D.B., Shu, Y., Guilinger, J.P., Bessen, J.L., Hu, J.H., Maeder, M.L., Joung, J.K., Chen, Z.-Y., and Liu, D.R. (2015). Cationic lipid-mediated delivery of proteins enables efficient protein-based genome editing in vitro and in vivo. Nat. Biotechnol. *33*, 73–80. http://dx.doi.org/10.1038/nbt.3081

CRISPR-mediated Genome Engineering and its Application in Industry

7

Saeed Kaboli[1]* and Hasan Babazada[2]

[1]Department of Biotechnology, Graduate School of Engineering, Osaka University, Suita, Japan.
[2]Department of Anesthesiology and Critical Care, Perelman School of Medicine, University of Pennsylvania, PA, USA.

*Correspondence: kaboli2009@gmail.com

https://doi.org/10.21775/9781910190630.07

Abstract

The CRISPR (clustered regularly interspaced short palindromic repeat)/Cas9 (CRISPR-associated nuclease 9) method has been dramatically changing the field of genome engineering. It is a rapid, highly efficient and versatile tool for precise modification of genome that uses a guide RNA (gRNA) to target Cas9 to a specific sequence. This novel RNA-guided genome-editing technique has become a revolutionary tool in biomedical science and has many innovative applications in different fields. In this review, we briefly introduce the Cas9-mediated genome-editing tool, summarize the recent advances in CRISPR/Cas9 technology to engineer the genomes of a wide variety of organisms, and discuss their applications to treatment of fungal and viral disease. We also discuss advantageous of CRISPR/Cas9 technology to drug design, creation of animal model, and to food, agricultural and energy sciences. Adoption of the CRISPR/Cas9 technology in biomedical and biotechnological researches would create innovative applications of it not only for breeding of strains exhibiting desired traits for specific industrial and medical applications, but also for investigation of genome function.

Introduction

Although in 1987 a group of scientists led by Atsuo Nakata (Osaka University) reported an unusual pattern of non-coding DNA in *Escherichia coli* (Ishino *et al.*, 1987), biological function of CRISPR arrays was not understood until 2005, when for the first time three studies suggested a role of it in adaptive immunity (Bolotin *et al.*, 2005; Mojica *et al.*, 2005; Pourcel *et al.*, 2005). Then, in 2007, Barrangou *et al.* (2007) provided evidence of adaptive immunity in bacteria by monitoring clustered regularly interspaced short palindromic repeats (CRISPR) loci in phage-challenged cultures of *Streptococcus thermophilus* (one type of bacteria used to make yogurt and cheese). In addition, Horvath's research group reported that bacteria harbouring a particular viral sequence as a CRISPR spacer were resistant to that virus, and that the CRISPR arrays were certified to provide protection against invading viruses when combined with Cas genes (Doudna and Charpentier, 2014). The mechanism of this immune system based on RNA-mediated DNA targeting was illustrated shortly thereafter (Brouns *et al.*, 2008; Deltcheva *et al.*, 2011; Garneau *et al.*, 2010; Marraffini and Sontheimer, 2008). In 2012, a research team led by Emmanuelle Charpentier and Jennifer Doudna devised the Type II CRISPR system from *Streptococcus pyogenes* for genome editing (Jinek *et al.*, 2012). Consequently, researchers can now utilize this mechanism to break the genome of most organisms, prokaryotic or eukaryotic, at almost any site (Fig. 7.1). The system requires only two components: the Cas9 DNA endonuclease and a single guide RNA (sgRNA) encoding the reverse complement to the sequence in the DNA to be targeted.

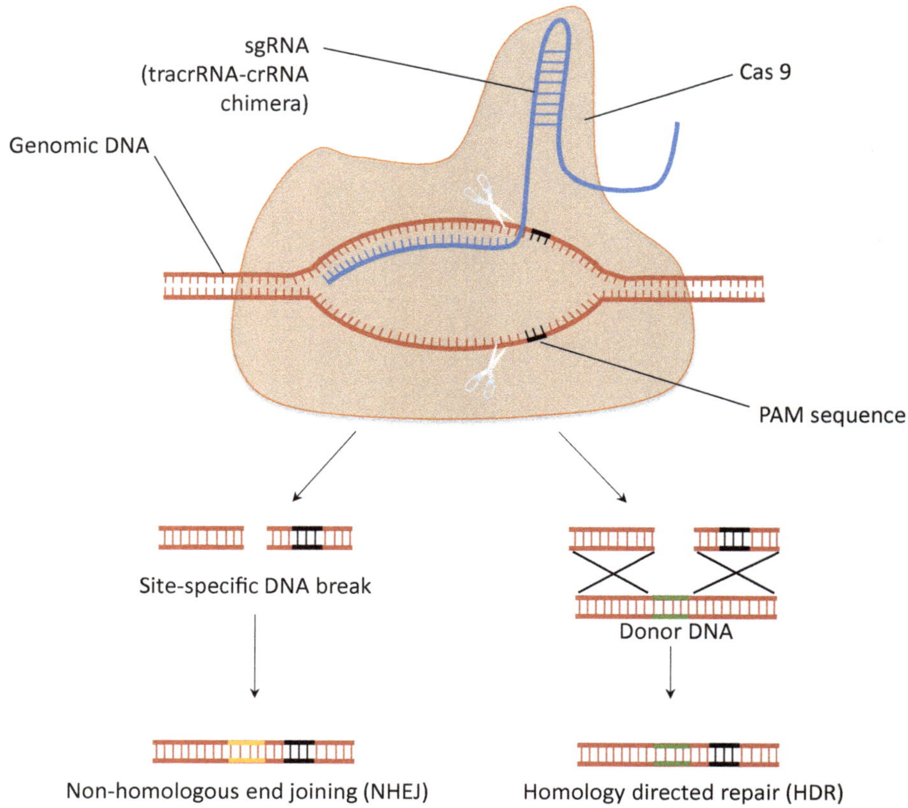

Figure 7.1 The CRISPR/Cas system. The CRISPR-associated endonuclease Cas9 could target specific DNA loci and make double-strand breaks (DSBs) under the guidance of the sgRNA. The presence of the Protospacer Adjacent Motif (PAM) at the target sequence is mandatory for successful Cas9 binding and catalysis – a protection mechanism to avoid self-cutting in CRISPR-containing organisms. DNA double strand breaks can be repaired by homologous recombination (HR) or non-homologous end joining (NHEJ). NHEJ re-ligates DSBs in an error-prone manner allowing for insertion or deletion of several bases at the DSB site. HR, on the hand, is a high-fidelity repair mechanism that uses an identical (or very similar) copy of the DSB region as a template for repair.

The only restriction for designing a CRISPR/Cas9 guiding sequence is a need for a protospacer adjacent motif (PAM) close to genomic target site (Jinek *et al.*, 2012; Cong *et al.*, 2013). Unlike ZFNs and TALENs, which achieve sequence recognition via protein–DNA interactions, Cas9 can be targeted to specific genomic loci with a guide RNA (gRNA). In addition, ZFNs and TALENs are costly, less reliable and time-consuming techniques than CRISPR/Cas9 for research. In eukaryotes, once the Cas9–sgRNA complex introduces a Double-Strand Break (DSB) at the target site, it is immediately repaired through the evolutionarily conserved pathways of error-prone non-homologous end joining (NHEJ) (which consists of re-ligation of the DSB resulting in loss or addition of a few nucleotides) or homology-directed repair (HDR) (when cells harbour a second copy of chromosomes) (Fig. 7.1) (Barnes, 2001; Van den Bosch *et al.*, 2002). With the utilization of these repair processes, researchers have been able to disrupt specific genes, add exogenous DNA elements into intended genomic sites, introduce single-nucleotide substitutions, and perform many other applications.

Despite slight improvement by using traditional molecular biology tools, most organisms relevant to biomedical and biotechnological sciences still show resistance to genetic engineering which could be problematic from the viewpoint of breeding new energy producing industrial organisms. To date,

Table 7.1 Application of CRISPR/Cas9 to industrially important organisms

Organism	Cell type	Industry	Reference
Fungi			
Saccharomyces cerevisiae	CEN.PK	Energy	Ronda *et al.*, 2015
Candida albicans	SC5314	Health care	Vyas *et al.*, 2015
	SC5314, BWP17, SN152	Health care	Min *et al.*, 2016
Trichoderma reesei	Qm6a	Chemical	Liu *et al.*, 2015
Aspergillus oryzae		Brewing	Katayama *et al.*, 2016
Yarrowia lipolytica	NS18 and NS432	Energy	Friedlander *et al.*, 2016
Streptomyces coelicolor	A3(2)	Energy	Tong *et al.*, 2015
Bacteria			
Escherichia coli	BL21 and BW25113-T7	Health care	Ahmed *et al.*, 2014
	IYB5670 and IYB5671	Health care	Yosef *et al.*, 2015
	DH5α	Health care	Stovicek *et al.*, 2015
Clostridium autoethanogenum	DSM10061	Energy	Nagaraju *et al.*, 2016
Virus			
HIV-1	T cells	Health care	Ebina *et al.*, 2013
	Microglia, promonocytes, T cells	Health care	Hu *et al.*, 2014a
HPV18	HeLa	Health care	Kennedy *et al.*, 2014
HPV16	SiHa	Health care	Kennedy *et al.*, 2014
HPV16	SiHa and CaSki	Health care	Hu *et al.*, 2014b
HBV	HepAD38	Health care	Kennedy *et al.*, 2015
HBV	HepG2.2.15	Health care	Zhen *et al.*, 2015
EBV	Raji	Health care	Wang and Quake, 2014

CRISPR/Cas9 has been used to overcome the challenges with these organisms through its application for treatment of microbial disease (bacterial, fungal and viral), for generation of animal disease models and drug discovery, and for improvement of food, agricultural and energy sciences. However, one outstanding limitation to the technique is off-target mutations, in which Cas9-directed genome editing occurs at unintended DNA sites (Jinek *et al.*, 2012; Cong *et al.*, 2013; Fu *et al.*, 2013). Although current works represent that these objections are not a major difficulty, precise qualification of Cas9 is needed before direct usage of the technology in humans. Ethical concerns are a second leading issue (Webber, 2014). There are concerns about the risk of errors in heritable modifications and (unintended) consequences. Moreover, *S. pyogenes* Cas9 is the most widely used orthologue for genome editing and requires an NGG (PAM) (Jiang *et al.*, 2013). Each of the GG dinucleotides is present on average in 1 of every 8 base pairs; therefore, extended genomic regions without PAM are very rare in most organisms (Cong *et al.*, 2013).

Application of CRISPR technology to treatment of microbial infection

Application of CRISPR technology to the treatment of fungal infection

Mycosis is a fungal infection of animals and humans (Kirk *et al.*, 2008). Mycoses are common and a variety of environmental and physiological conditions can contribute to the development of fungal diseases. Fighting mycosis is an increasingly important global health concern and involves dominating both emerging infectious agents and newly drug-resistant strains. Among microbial infection, mycosis accounts for more deaths annually than even either tuberculosis or malaria (Denning and Bromley, 2015). Therefore, new biological leads for pathogen identification and therapeutics are required, and researchers have developed new genome engineering tools for the study of pathogenic fungi such as those that are recombination-based techniques (Krappmann *et al.*, 2006; Takahashi *et al.*, 2008). By using these techniques, a major limitation for functional analysis of genome in fungal pathogens has been the low rates of homologous recombination between exogenous DNA sequences and recipient genomes. On the other hand, because there are a limited number of validated selectable markers for transformation and making knockouts of multiple genes, it is impractical for large numbers of genes to be analysed in a single-strain background. To circumvent these problems several site-specific recombinase systems have been developed (Hartmann *et al.*, 2010; Shahana *et al.*, 2014; Khrunyk *et al.*, 2010). However, among all genome editing tools the CRISPR/Cas9 system has been exhibiting to be a rapid and facile genome manipulation for both yeasts (Ronda *et al.*, 2015; Laughery *et al.*, 2015; DiCarlo *et al.*, 2013; Vyas *et al.*, 2015; Min *et al.*, 2016) and filamentous fungi (Liu *et al.*, 2015; Arazoe *et al.*, 2015; Nødvig *et al.*, 2015; Fuller *et al.*, 2015; Zhang *et al.*, 2016; Katayama *et al.*, 2015; Matsu-ura *et al.*, 2015). For example, *Candida albicans* is pathogenic yeast that causes mucosal and systematic infections with high mortality. Gerald R. Fink's group used the CRISPR system to create homozygous gene knockouts, mutations in multiple genes and genes that encode essential function. The ability to analyse essential genes provides an opportunity to explore potential antifungal targets in *C. albicans* (Vyas *et al.*, 2015). Their finding that CRISPR works effectively in a recent antifungal resistant clinical isolate suggests a new route to characterize clinical isolates of drug-resistant strains of *Candida*.

Application of CRISPR technology to the treatment of viral infection

Viral infections are of therapeutic challenge since viral life cycles occur within the host cells. Recent years have seen the rapid development of novel application for genetic-engineering technologies including those for the treatment of viral infections. Studies have shown that the CRISPR/Cas9 system can clear the HIV-1 genome and prevent new HIV infection (Ebina *et al.*, 2013; Hu *et al.*, 2014a). In fact, sgRNA expression vector targeting the long terminal repeats (LTR) of HIV-1 efficiently cleaves and mutates LTR target sites and suppresses LTR-driven viral gene expression. In addition, Ebina group showed that this system can delete viral genes from the host cell chromosome (Ebina *et al.*, 2013). The high specificity of Cas9/sgRNAs in editing the HIV-1 target genome has also been recently demonstrated (Hu *et al.*, 2014a). Cas9/sgRNAs efficiently inactivate HIV gene expression and replication in latently infected cells, including microglial, promonocytic and T cells. Significantly, Cas9/sgRNA mediated genome editing has been shown to immunize cells to prevent HIV-1 infection (Hu *et al.*, 2014a). These results indicate that the CRISPR/Cas9 technology can serve as a potential tool for clinical applications to cure viral-based infectious diseases. CRISPR system has also been exhibited the potential to be developed as an effective therapy for human papillomavirus (HPV)-associated tumours (Kennedy *et al.*, 2014; Hu *et al.*, 2014b), hepatitis B virus (HBV)-associated diseases (Kennedy *et al.*, 2015; Zhen *et al.*, 2015) and Epstein–Barr virus (EBV) (Wang and Quake, 2014) in the clinic.

Application of CRISPR technology to drug design

So far, drug design investments have not been returned as medicines and budgets are being cut (Paul *et al.*, 2010), partly because some clinical failures happen due to molecule quality and in some cases therapeutic hypothesis is without foundation (Arrowsmith, 2011; Prinz *et al.*, 2011). However, the abilities to apply biomedical knowledge in drug discovery have been improving; availability of knock-out and knock-in mice has helped model generation (Heck, 2004) and RNA interference (RNAi) has aided target validation (Bartz and Jackson, 2005). However, RNAi has its drawbacks, including partial knock-down that can fail to produce a measurable phenotype (Sachse *et al.*, 2005) and off-target effects (Marine *et al.*, 2012). Despite these drawbacks, both short hairpin RNA (shRNA) and small interfering RNA (siRNA) genome-wide screens have been used to look for new therapeutic targets in a number of diseases.

The discovery of the Cas9 nuclease removed many of the technical and financial barriers to high-throughput knock-out screens (Malina *et al.*, 2013; Wang *et al.*, 2014; Shalem *et al.*, 2014; Koike-Yusa *et al.*, 2014; Zhou *et al.*, 2014; Sanjana *et al.*, 2014; Doench *et al.*, 2014): specific endonucleases can be assembled in cells simply by co-expressing Cas9 and an sgRNA, which is short enough to be encoded by oligonucleotides that can be synthesized in large-scale arrays. Antibiotic overuse in medicine and abuse in animal agriculture has led to the rise of multidrug-resistant pathogens that are increasingly tolerant to the current antibiotic arsenal. Accordingly, there is a need for novel antimicrobials that can bypass common modes of multidrug resistance while being selective for individual strains. Because CRISPR/Cas9 is a more precise and a sequence-specific technique than previous similar ones, it is possible to use its specificity to design it to target a single bacterial species. It will only cut up essential genes from that one species, even when the target species is mixed up with others. Therefore, antimicrobial CRISPR/Cas systems may be better weapons against bacteria than antibiotics. In this regard, the Marraffini group in their work suggested that CRISPR/Cas systems could be used for the sequence-specific killing of bacteria (Bikard *et al.*, 2012). Later studies by Bikard *et al.* (2014), Citorik *et al.* (2014) and Gomaa *et al.* (2014) offered a promising solution to the problem of antibiotic resistance by using CRISPR/Cas9 system. Yosef *et al.* (2015) also designed a CRISPR-based system that could selectively kill antibiotic-resistant bacteria. They used temperate phages to deliver a functional CRISPR/Cas system into the genome of antibiotic-resistant bacteria. The delivered CRISPR/Cas system destroyed both antibiotic resistance-conferring plasmids and genetically modified lytic phages. With further development, CRISPR has the potential to treat multidrug-resistant infections without impacting beneficial microbes, to remove contaminating microbes from industrial fermentations (Stovicek *et al.*, 2015) and to provide further insights into microbial communities.

Application of CRISPR technology to generation of animal models

Cas9-mediated genome editing has enabled accelerated generation of transgenic models and expands biological research beyond traditional, genetically tractable animal model organisms (Sander and Joung, 2014). For generation of cellular models, Cas9 can be easily introduced into the target cells

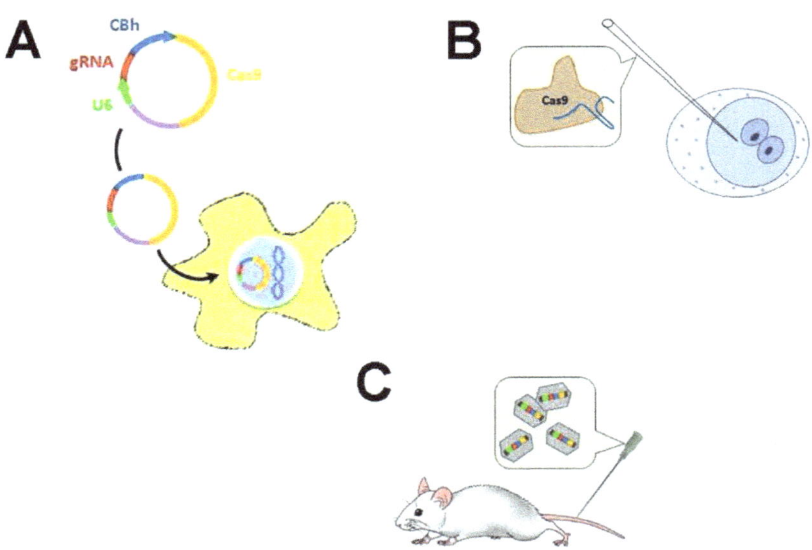

Figure 7.2 Applications of Cas9. (A) Expression plasmids encoding both the Cas9 gene and a short sgRNA cassette driven by the U6 RNA polymerase III promoter can be directly transfected into any cell line of interest. (B) Purified Cas9 protein and *in vitro* transcribed sgRNA can be microinjected into fertilized zygotes for rapid generation of transgenic animal models. (C) In order to perform somatic genetic modification, high-titre viral vectors encoding CRISPR reagents can be transduced into tissues or cells of interest.

using transient transfection of plasmids carrying Cas9 and the appropriately designed sgRNA (Fig. 7.2A). Additionally, the multiplexing capabilities of Cas9 offer a promising approach for studying common human diseases – such as diabetes, heart disease, schizophrenia, and autism – that are typically polygenic. For generation of transgenic animal models, Cas9 protein and transcribed sgRNA can be directly injected into fertilized zygotes to achieve heritable gene modification at one or multiple alleles in models such as rodents and monkeys (Wang *et al.*, 2013; Li *et al.*, 2013; Yang *et al.*, 2013; Niu *et al.*, 2014) (Fig. 7.2B). Successful multiplex targeting in cynomolgus monkey models was also recently reported (Niu *et al.*, 2014), suggesting the potential for establishing more accurate modelling of complex human diseases such as neuropsychiatric disorders using primate models. Additionally, Cas9 could be harnessed for direct modification of somatic tissue, obviating the need for embryonic manipulation (Fig. 7.2C) as well as enabling therapeutic use for gene therapy.

Studies to date have typically relied on the injection of Cas9 mRNA into zygotes (fertilized embryos at the single-cell stage). However, because transcription and translation activity is suppressed in the mouse zygote, Cas9 mRNA translation into active enzymatic form is likely delayed until after the first cell division (Oh *et al.*, 2000). Because NHEJ-mediated repair is thought to introduce indels of random length, this translation delay likely plays a major role in contributing to genetic mosaicism in CRISPR-modified mice. To overcome this limitation, Cas9 protein and sgRNA could be directly injected into single-cell fertilized embryos. The high rate of non-mutagenic repair by the NHEJ process may additionally contribute to undesired mosaicism because introducing indels that mutate the Cas9 recognition site would then have to compete with zygotic division rates. To increase the mutagenic activity of NHEJ, a pair of sgRNAs flanking a small fragment of the target gene may be used to increase the probability of gene disruption. Tan and colleagues (Tan *et al.*, 2013) used CRISPR system to produce biomedical model pigs to enhance productivity in the livestock industry. Xue *et al.* (2014) used hydrodynamic injection to deliver plasmids encoding sgRNA and Cas9 to the murine liver where simultaneous targeting of the Pten and Trp53 led to the development of liver tumours in all five mice treated within 3 months. Platt *et al.* (2014) generated a Cre-dependent Cas9 knock-in mouse, which could be crossed with a variety of Cre-driver strains to enable the expression of Cas9 either constitutively or in various tissues. Multiple reports describe the use of custom nucleases to engineer cultured human stem cells. Cas9-sgRNA and a plasmid donor were employed to correct the cystic fibrosis transmembrane conductance receptor by homologous recombination in cultured intestinal stem cells from cystic fibrosis patients: organoids derived from these cells restored the normal cAMP-induced swelling phenotype (Schwank *et al.*, 2013).

Application of CRISPR technology to food and agricultural sciences

One of the major goals in the food and agricultural sciences is to get technologies to develop health-promoting products for a growing world population. In this regard, the use of recombinant genetic technologies has profoundly impacted molecular biology research and applications in fields such as biosynthesis of vitamins, enzymes, pharmaceuticals, antibiotics, and bioactive peptides. The CRISPR/Cas9 technology is a modern, fashionable method in plant research. Immediately after its early use to edit the genomes of animals and bacteria (Hwang *et al.*, 2013; Jiang *et al.*, 2013; Mali *et al.*, 2013), in August 2013, five reports were published discussing the first application of CRISPR/Cas9-based genome editing in plants (Feng *et al.*, 2013; Li *et al.*, 2013; Nekrasov *et al.*, 2013; Shan *et al.*, 2013; Xie and Yang, 2013). Subsequent work focused on additional crop species such as sorghum (Jiang *et al.*, 2013), wheat (Upadhyay *et al.*, 2013; Wang *et al.*, 2014) and maize (Liang *et al.*, 2014). Two research groups simultaneously proposed a possible strategy to make plants virus resistant using CRISPR/Cas9 technology (Baltes *et al.*, 2015; Ji *et al.*, 2015). In addition, the application of CRISPR/Cas9 could extend to a direct knock-out strategy in the disease susceptible genes, often termed 'S-genes', of a host genome; the result is the development of durable disease-resistant crops. The S-gene knock-out strategy was employed successfully in hexaploid bread wheat using the TALEN and CRISPR/Cas9 system, and the fungi-resistant wheat was generated (Wang *et al.*, 2014). Thus, novel and valuable plants generated by CRISPR/Cas9 can regain useful traits overlooked during domestication; these traits help plants survive unpredictable global environmental changes.

Application of CRISPR technology to energy sciences

Because CRISPR/Cas9 has been shown to lead to precise and affordable genome edition in bioenergetics systems, then it can be very powerful tool in energy sciences. *Yarrowia lipolytica* is valuable oleaginous microbial host for chemical production known for converting sugars to lipids and hydrocarbons as energy source that are difficult to make synthetically (Friedlander *et al.*, 2016). Genome of the *Y. lipolytica* has been hard to manipulate at the genetic level by old genome engineering tools. In a recent work, however, a research team could adapt CRISPR/Cas9 system for *Y. lipolytica*, showing that the system could be used for markerless gene knock-out and introduction of new genes (Schwartz *et al.*, 2016). This approach, based on the team's claim, is the first step of a project to create long chain hydrocarbons – used to make specialty polymers, adhesives, coatings and fragrances – from yeast rather than synthetically.

Bacteria are almost unlimited source of enzymes, and they are extensively used in industry in various ways for the manufacture of dairy products, and the production of biological substances such as enzymes, vaccines, antibiotics and biofuels. In this regard, *Clostridium autoethanogenum* is a model acetogen that is being pursued for fuel (ethanol) and chemical (2,3-butanediol) production at commercial scale (Liew *et al.*, 2016; Daniell *et al.*, 2016). Recently, Nagaraju *et al.* (2016) constructed and screened a small library of tetracycline-inducible promoters in CRISPR/Cas9 system and reported that they could improve the efficiency of CRISPR/Cas9-mediated desired gene deletion to over 50%, making a viable tool for engineering *C. autoethanogenum*.

Sustainable and cost-effective biofuels are attractive sources for renewable energy. This could be achieved by either creating of efficient metabolic pathways for ethanol production in organisms such as algae or by engineering of yeast, mould and bacteria which have shown the potential to generate biofuel from plant biomasses like maize biomass. Despite some improvement, most of these organisms show resistant to engineer by traditional molecular biology tools which could be problematic from the viewpoint of breeding new energy producing industrial strains. CRISPR/Cas9 has been used to overcome those genome editing drawbacks and limitations with organisms (DiCarlo *et al.*, 2013; Min *et al.*, 2016; Nødvig *et al.*, 2015; Matsu-ura *et al.*, 2015; Stovicek *et al.*, 2015; Shan *et al.*, 2013; Mans *et al.*, 2015; Jacobs *et al.*, 2014; Ninomiya *et al.*, 2004; Huang *et al.*, 2015; Tong *et al.*, 2015; Kim *et al.*, 2015; Feng *et al.*, 2016).

Future direction

Cas9-based technology is becoming an ultimate molecular powerful tool for studies in basic, biomedical and biotechnological sciences. This technology has been successfully applied for genetic manipulation in numerous bacterial, fungal, viral and other species. Adapting CRISPR/Cas to industrially and medically important organisms is highly desirable because they are the central core in energy, pharmaceutical and health care industries. Therefore, for those host organisms a better characterization of genetics and subsequently a more validated Cas9-based engineering package are essential (Crook and Alper, 2012). In this way, one major obstacle might be off-target effects and their consequences. In the future, the use of high-throughput methods that enable comprehensive profiling of off–target cleavage sites (Jinek *et al.*, 2012; Cong *et al.*, 2013; Fu *et al.*, 2013) should provide low-cost and high-speed engineering in each host system. In case of ethical issues as one of the increasingly important concerns, the possibility of modifying the germline – in particular the human germline – should be considered carefully.

Acknowledgements

This work was partially supported by the Global COE (centre of excellence) Program 'Global Education and Research Center for Bio-Environmental Chemistry' of Osaka University.

Conflict of interest statement

There is no conflict.

References

Arazoe, T., Miyoshi, K., Yamato, T., Ogawa, T., Ohsato, S., Arie, T., and Kuwata, S. (2015). Tailor-made CRISPR/Cas system for highly efficient targeted gene replacement in the rice blast fungus. Biotechnol. Bioeng. *112*, 2543–2549. http://dx.doi.org/10.1002/bit.25662

Arrowsmith, J. (2011). Trial watch: Phase II failures: 2008-2010. Nat. Rev. Drug Discovery *10*, 328–329. http://dx.doi.org/10.1038/nrd3439

Baltes, N.J., Hummel, A.W., Konecna, E., Cegan, R., Bruns, A.N., Bisaro, D.M., and Voytas, D.F. (2015). Conferring resistance to geminiviruses with the CRISPR–Cas prokaryotic immune system. Nat. Plants. *1*, 15145. http://dx.doi.org/10.1038/nplants.2015.145

Barnes, D.E. (2001). Non-homologous end joining as a mechanism of DNA repair. Curr. Biol. *11*, R455–7.

Barrangou, R., Fremaux, C., Deveau, H., Richards, M., Boyaval, P., Moineau, S., Romero, D.A., and Horvath, P. (2007). CRISPR provides acquired resistance against viruses in prokaryotes. Science *315*, 1709–1712.

Bartz, S., and Jackson, A.L. (2005). How will RNAi facilitate drug development? Sci. STKE. *2005*, pe39.

Bikard, D., Euler, C.W., Jiang, W., Nussenzweig, P.M., Goldberg, G.W., Duportet, X., Fischetti, V.A., and Marraffini, L.A. (2014). Exploiting CRISPR-Cas nucleases to produce sequence-specific antimicrobials. Nat. Biotechnol. *32*, 1146–1150. http://dx.doi.org/10.1038/nbt.3043

Bikard, D., Hatoum-Aslan, A., Mucida, D., and Marraffini, L.A. (2012). CRISPR interference can prevent natural transformation and virulence acquisition during in vivo bacterial infection. Cell Host Microbe. *12*:177–186. http://dx.doi.org/10.1016/j.chom.2012.06.003

Bolotin, A., Quinquis, B., Sorokin, A., and Ehrlich, S.D. (2005). Clustered regularly interspaced short palindrome repeats (CRISPRs) have spacers of extrachromosomal origin. Microbiology. *151*, 2551–2561. http://dx.doi.org/10.1099/mic.0.28048–0

Brouns, S.J., Jore, M.M., Lundgren, M., Westra, E.R., Slijkhuis, R.J., Snijders, A.P., Dickman, M.J., Makarova, K.S., Koonin, E.V., and van der Oost, J. (2008). Small CRISPR RNAs guide antiviral defense in prokaryotes. Science *321*, 960–964. http://dx.doi.org/10.1126/science.1159689

Citorik, R.J., Mimee, M., and Lu, T.K. (2014). Sequence-specific antimicrobials using efficiently delivered RNA-guided nucleases. Nat. Biotechnol. *32*, 1141–1145. http://dx.doi.org/10.1038/nbt.3011

Cong, L., Ran, F.A., Cox, D., Lin, S., Barretto, R., Habib, N., Hsu, P.D., Wu, X., Jiang, W., Marraffini, L.A., *et al.* (2013). Multiplex genome engineering using CRISPR/Cas systems. Science *339*, 819–823. http://dx.doi.org/10.1126/science.1231143

Crook, N., and Alper, H.S. (2012). Classical strain improvement. Engineering complex phenotypes in industrial strains. (John Wiley & Sons, Inc.). http://dx.doi.org/10.1002/9781118433034.ch1

Daniell, J., Nagaraju, S., Burton, F., Köpke, M., and Simpson, S.D. (2016). Low-Carbon Fuel and Chemical Production by Anaerobic Gas Fermentation. Adv. Biochem. Eng. Biotechnol. *156*, 293–321. http://dx.doi.org/10.1007/10_2015_5005

Deltcheva, E., Chylinski, K., Sharma, C.M., Gonzales, K., Chao, Y., Pirzada, Z.A., Eckert, M.R., Vogel, J., and Charpentier, E. (2011). CRISPR RNA maturation by trans-encoded small RNA and host factor RNase III. Nature *471*, 602–607. http://dx.doi.org/10.1038/nature09886

Denning, D.W., and Bromley, M.J. (2015). Infectious disease. How to bolster the antifungal pipeline. Science *347*, 1414–1416. http://dx.doi.org/10.1126/science.aaa6097

DiCarlo, J.E., Norville, J.E., Mali, P., Rios, X., Aach, J., and Church, G.M. (2013). Genome engineering in *Saccharomyces cerevisiae* using CRISPR-Cas systems. Nucleic Acids Res. *41*, 4336–4343. http://dx.doi.org/10.1093/nar/gkt135

Doench, J.G., Hartenian, E., Graham, D.B., Tothova, Z., Hegde, M., Smith, I., Sullender, M., Ebert, B.L., Xavier, R.J., and Root, D.E. (2014). Rational design of highly active sgRNAs for CRISPR-Cas9-mediated gene inactivation. Nat. Biotechnol. *32*, 1262–1267. http://dx.doi.org/10.1038/nbt.3026

Doudna, J.A., and Charpentier, E. (2014). Genome editing. The new frontier of genome engineering with CRISPR-Cas9. Science *346*, 1258096. http://dx.doi.org/10.1126/science.1258096

Ebina, H., Misawa, N., Kanemura, Y., and Koyanagi, Y. (2013). Harnessing the CRISPR/Cas9 system to disrupt latent HIV-1 provirus. Sci. Rep. *3*, 2510. http://dx.doi.org/10.1038/srep02510

Feng, C., Yuan, J., Wang, R., Liu, Y., Birchler, J.A., and Han, F. (2016). Efficient targeted genome modification in maize using CRISPR/Cas9 system. J. Genet. Genomics. *43*, 37–43. http://dx.doi.org/10.1016/j.jgg.2015.10.002

Feng, Z., Zhang, B., Ding, W., Liu, X., Yang, D.L., Wei, P., Cao, F., Zhu, S., Zhang, F., Mao, Y., *et al.* (2013). Efficient genome editing in plants using a CRISPR/Cas system. Cell Res. *23*, 1229–1232. http://dx.doi.org/10.1038/cr.2013.114

Friedlander, J., Tsakraklides, V., Kamineni, A., Greenhagen, E.H., Consiglio, A.L., MacEwen, K., Crabtree, D.V., Afshar, J., Nugent, R.L., Hamilton, M.A., *et al.* (2016). Engineering of a high lipid producing *Yarrowia lipolytica* strain. Biotechnol. Biofuels *9*, 77. http://dx.doi.org/10.1186/s13068-016-0492-3

Fu, Y., Foden, J.A., Khayter, C., Maeder, M.L., Reyon, D., Joung, J.K., and Sander, J.D. (2013). High-frequency off-target mutagenesis induced by CRISPR-Cas nucleases in human cells. Nat. Biotechnol. *31*, 822–826. http://dx.doi.org/10.1038/nbt.2623

Fuller, K.K., Chen, S., Loros, J.J., and Dunlap, J.C. (2015). Development of the CRISPR/Cas9 System for Targeted Gene Disruption in *Aspergillus fumigatus*. Eukaryot. Cell. *14*, 1073–1080. http://dx.doi.org/10.1128/EC.00107-15

Garneau, J.E., Dupuis, M.È., Villion, M., Romero, D.A., Barrangou, R., Boyaval, P., Fremaux, C., Horvath, P., Magadán, A.H., and Moineau, S. (2010). The CRISPR/Cas bacterial immune system cleaves bacteriophage and plasmid DNA. Nature *468*, 67–71. http://dx.doi.org/10.1038/nature09523

Gomaa, A.A., Klumpe, H.E., Luo, M.L., Selle, K., Barrangou, R., and Beisel, C.L. (2014). Programmable removal of bacterial strains by use of genome-targeting

CRISPR-Cas systems. mBio 5, e00928–13. http://dx.doi.org/10.1128/mBio.00928-13

Hartmann, T., Dümig, M., Jaber, B.M., Szewczyk, E., Olbermann, P., Morschhäuser, J., and Krappmann, S. (2010). Validation of a self-excising marker in the human pathogen *Aspergillus fumigatus* by employing the beta-rec/six site-specific recombination system. Appl. Environ. Microbiol. *76*, 6313–6317. http://dx.doi.org/10.1128/AEM.00882–10

Heck, S., Qian, X., and Velleca, M. (2004). Genetically engineered mouse models for drug discovery: new chemical genetic approaches. Curr. Drug. Discov. Technol. *1*, 13–26.

Hu, W., Kaminski, R., Yang, F., Zhang, Y., Cosentino, L., Li, F., Luo, B., Alvarez-Carbonell, D., Garcia-Mesa, Y., Karn, J., *et al.* (2014a). RNA-directed gene editing specifically eradicates latent and prevents new HIV-1 infection. Proc. Natl. Acad. Sci. U.S.A. *111*, 11461–11466. http://dx.doi.org/10.1073/pnas.1405186111

Hu, Z., Yu, L., Zhu, D., Ding, W., Wang, X., Zhang, C., Wang, L., Jiang, X., Shen, H., He, D., et al. (2014b). Disruption of HPV16-E7 by CRISPR/Cas system induces apoptosis and growth inhibition in HPV16 positive human cervical cancer cells. BioMed. Research International. ID 612823. http://dx.doi.org/10.1155/2014/612823

Huang, H., Zheng, G., Jiang, W., Hu, H., and Lu, Y. (2015). One-step high-efficiency CRISPR/Cas9-mediated genome editing in *Streptomyces*. Acta Biochim. Biophys. Sin. *47*, 231–243. http://dx.doi.org/10.1093/abbs/gmv007

Hwang, W.Y., Fu, Y., Reyon, D., Maeder, M.L., Tsai, S.Q., Sander, J.D., Peterson, R.T., Yeh, J.R., and Joung, J.K. (2013). Efficient genome editing in zebrafish using a CRISPR-Cas system. Nat. Biotechnol. *31*, 227–229. http://dx.doi.org/10.1038/nbt.2501

Ishino, Y., Shinagawa, H., Makino, K., Amemura, M., and Nakata, A. (1987). Nucleotide sequence of the iap gene, responsible for alkaline phosphatase isozyme conversion in *Escherichia coli*, and identification of the gene product. J. Bacteriol. *169*, 5429–5433.

Jacobs, J.Z., Ciccaglione, K.M., Tournier, V., and Zaratiegui, M. (2014). Implementation of the CRISPR-Cas9 system in fission yeast. Nat. Commun. *5*, 5344. http://dx.doi.org/10.1038/ncomms6344

Ji, X., Zhang, H., Zhang, Y., Wang, Y., and Gao, C. (2015). Establishing a CRISPR-Cas-like immune system conferring DNA virus resistance in plants. Nat. Plants *1*, 15144. http://dx.doi.org/10.1038/nplants.2015.144

Jiang, W., Bikard, D., Cox, D., Zhang, F., and Marraffini, L.A. (2013). RNA-guided editing of bacterial genomes using CRISPR-Cas systems. Nat. Biotechnol. *31*, 233–239. http://dx.doi.org/10.1038/nbt.2508

Jiang, W., Zhou, H., Bi, H., Fromm, M., Yang, B., and Weeks, D.P. (2013). Demonstration of CRISPR/Cas9/sgRNA-mediated targeted gene modification in *Arabidopsis*, tobacco, sorghum and rice. Nucleic Acids Res. *41*, e188. http://dx.doi.org/10.1093/nar/gkt780

Jinek, M., Chylinski, K., Fonfara, I., Hauer, M., Doudna, J.A., and Charpentier, E. (2012). A programmable dual-RNA-guided DNA endonuclease in adaptive bacterial immunity. Science *337*, 816–821. http://dx.doi.org/10.1126/science.1225829

Katayama, T., Tanaka, Y., Okabe, T., Nakamura, H., Fujii, W., Kitamoto, K., and Maruyama, J.-I. (2015). Development of a genome editing technique using the CRISPR/Cas9 system in the industrial filamentous fungus *Aspergillus oryzae*. Biotechnol. Lett. http://dx.doi.org/10.1007/s10529-015-2015-x

Kennedy, E.M., Bassit, L.C., Mueller, H., Kornepati, A.V., Bogerd, H.P., Nie, T., Chatterjee, P., Javanbakht, H., Schinazi, R.F., and Cullen, B.R. (2015). Suppression of hepatitis B virus DNA accumulation in chronically infected cells using a bacterial CRISPR/Cas RNA-guided DNA endonuclease. Virology. *476*, 196–205. http://dx.doi.org/10.1016/j.virol.2014.12.001

Kennedy, E.M., Kornepati, A.V., Goldstein, M., Bogerd, H.P., Poling, B.C., Whisnant, A.W., Kastan, M.B., and Cullen, B.R. (2014). Inactivation of the human papillomavirus E6 or E7 gene in cervical carcinoma cells by using a bacterial CRISPR/Cas RNA-guided endonuclease. J. Virol. *88*, 11965–11972. http://dx.doi.org/10.1128/JVI.01879-14

Khrunyk, Y., Münch, K., Schipper, K., Lupas, A.N., and Kahmann, R. (2010). The use of FLP-mediated recombination for the functional analysis of an effector gene family in the biotrophic smut fungus *Ustilago maydis*. New. Phytol. *187*, 957–968. http://dx.doi.org/10.1111/j.1469-8137.2010.03413.x

Kim, H., Kim, S.-T., Kim, S.-G., and Kim, J.-S. (2015). Targeted genome editing for crop improvement. Plant Breed. Biotech. 3, 283–290. http://dx.doi.org/10.9787/PBB.2015.3.4.283

Kirk, P.M., Cannon, P.F., Minter, D.W., and Stalpers, J.A. (2008). Dictionary of the fungi (CABI Europe – UK, Egham, and CBS, Utrecht).

Koike-Yusa, H., Li, Y., Tan, E.P., Velasco-Herrera, M.d.e.l.C., and Yusa, K. (2014). Genome-wide recessive genetic screening in mammalian cells with a lentiviral CRISPR-guide RNA library. Nat. Biotechnol. *32*, 267–273. http://dx.doi.org/10.1038/nbt.2800

Krappmann, S., Sasse, C., and Braus, G.H. (2006). Gene targeting in Aspergillus fumigatus by homologous recombination is facilitated in a nonhomologous end-joining-deficient genetic background. Eukaryot. Cell. *5*, 212–215.

Laughery, M.F., Hunter, T., Brown, A., Hoopes, J., Ostbye, T., Shumaker, T., and Wyrick, J.J. (2015). New vectors for simple and streamlined CRISPR-Cas9 genome editing in *Saccharomyces cerevisiae*. Yeast. *32*, 711–720. http://dx.doi.org/10.1002/yea.3098. Epub 2015 Sep 21

Li, J.-F., Norville, J.E., Aach, J., McCormack, M., Zhang, D., Bush, J., Church, G.M., and Sheen, J. (2013). Multiplex and homologous recombination-mediated genome editing in *Arabidopsis* and *Nicotiana benthamiana* using guide RNA and Cas9. Nat. Biotechnol. *31*, 688–691. http://dx.doi.org/10.1038/nbt.2654

Li, W., Teng, F., Li, T., and Zhou, Q. (2013). Simultaneous generation and germline transmission of multiple gene mutations in rat using CRISPR-Cas systems. Nat. Biotechnol. *31*, 684–686. http://dx.doi.org/10.1038/nbt.2652

Liang, Z., Zhang, K., Chen, K., and Gao, C. (2014). Targeted mutagenesis in Zea *mays* using TALENs and the CRISPR/Cas system. J. Genet. Genomics *41*, 63–68. http://dx.doi.org/10.1016/j.jgg.2013.12.001

Liew, F., Martin, M.E., Tappel, R.C., Heijstra, B.D., Mihalcea, C., and Köpke, M. (2016). Gas fermentation – a flexible platform for commercial scale production of low-carbon-fuels and chemicals from waste and renewable feedstocks. Front. Microbiol. 7, 694. http://dx.doi.org/10.3389/fmicb.2016.00694

Liu, R., Chen, L., Jiang, Y., Zhou, Z., and Zou, G. (2015). Efficient genome editing in filamentous fungus *Trichoderma reesei* using the CRISPR/Cas9 system. Cell Discovery *1*, 15007. http://dx.doi.org/10.1038/celldisc.2015.7

Mali, P., Yang, L., Esvelt, K.M., Aach, J., Guell, M., DiCarlo, J.E., Norville, J.E., and Church, G.M. (2013). RNA-guided human genome engineering via Cas9. Science *339*, 823–826. http://dx.doi.org/10.1126/science.1232033

Malina, A., Mills, J.R., Cencic, R., Yan, Y., Fraser, J., Schippers, L.M., Paquet, M., Dostie, J., and Pelletier, J. (2013). Repurposing CRISPR/Cas9 for in situ functional assays. Genes. Dev. *27*, 2602–2614. http://dx.doi.org/10.1101/gad.227132.113

Mans, R., van Rossum, H.M., Wijsman, M., Backx, A., Kuijpers, N.G., van den Broek, M., Daran-Lapujade, P., Pronk, J.T., van Maris, A.J., and Daran, J.M. (2015). CRISPR/Cas9: a molecular Swiss army knife for simultaneous introduction of multiple genetic modifications in *Saccharomyces cerevisiae*. FEMS Yeast Res. *15*, fov004. http://dx.doi.org/10.1093/femsyr/fov004

Marine, S., Bahl, A., Ferrer, M., and Buehler, E. (2012). Common seed analysis to identify off-target effects in siRNA screens. J. Biomol. Screening *17*, 370–378. http://dx.doi.org/10.1177/1087057111427348

Marraffini, L.A., and Sontheimer, E.J. (2008). CRISPR interference limits horizontal gene transfer in *staphylococci* by targeting DNA. Science *322*, 1843–1845. http://dx.doi.org/10.1126/science.1165771

Matsu-ura, T., Baek, M., Kwon, J., and Hong, C. (2015). Efficient gene editing in *Neurospora crassa* with CRISPR technology. Fungal Biol. Biotechnol. 2, http://dx.doi.org/10.1186/s40694-015-0015-1

Min, K., Ichikawa, Y., Woolford, C.A., and Mitchell, A.P. (2016). *Candida albicans* Gene Deletion with a Transient CRISPR-Cas9 System. mSphere *1*, e00130–16. http://dx.doi.org/10.1128/mSphere.00130-16

Mojica, F.J., Diez-Villasenor, C., Garcia-Martinez, J., and Soria, E. (2005). Intervening sequences of regularly spaced prokaryotic repeats derive from foreign genetic elements. J. Mol. Evol. *60*, 174–182. http://dx.doi.org/10.1007/s00239-004-0046-3

Nagaraju, S., Davies, N.K., Walker, D.J., Köpke, M., and Simpson, S.D. (2016). Genome editing of *Clostridium autoethanogenum* using CRISPR/Cas9. Biotechnol. Biofuels *9*, 219. http://dx.doi.org/10.1186/s13068-016-0638-3

Nekrasov, V., Staskawicz, B., Weigel, D., Jones, J.D., and Kamoun, S. (2013). Targeted mutagenesis in the model plant *Nicotiana benthamiana* using Cas9 RNA-guided endonuclease. Nat. Biotechnol. *31*, 691–693. http://dx.doi.org/10.1038/nbt.2655

Ninomiya, Y., Suzuki, K., Ishii, C., and Inoue, H. (2004). Highly efficient gene replacements in *Neurospora* strains deficient for nonhomologous end-joining. Proc. Natl. Acad. Sci. U.S.A. *101*, 12248–12253. http://dx.doi.org/10.1073/pnas.0402780101

Niu, Y., Shen, B., Cui, Y., Chen, Y., Wang, J., Wang, L., Kang, Y., Zhao, X., Si, W., Li, W., *et al.* (2014). Generation of gene-modified cynomolgus monkey via Cas9/RNA-mediated gene targeting in one-cell embryos. Cell *156*, 836–843. http://dx.doi.org/10.1016/j.cell.2014.01.027

Nødvig, C.S., Nielsen, J.B., Kogle, M.E., and Mortensen, U.H. (2015). A CRISPR-Cas9 system for genetic engineering of filamentous fungi. PLOS One *10*, e0133085. http://dx.doi.org/10.1371/journal.pone.0133085

Oh, B., Hwang, S., McLaughlin, J., Solter, D., and Knowles, B.B. (2000). Timely translation during the mouse oocyte-to-embryo transition. Development *127*, 3795–3803.

Paul, S.M., Mytelka, D.S., Dunwiddie, C.T., Persinger, C.C., Munos, B.H., Lindborg, S.R., and Schacht, A.L. (2010). How to improve R&D productivity: the pharmaceutical industry's grand challenge. Nat. Rev. Drug Discovery *9*, 203–214. http://dx.doi.org/10.1038/nrd3078

Platt, R.J., Chen, S., Zhou, Y., Yim, M.J., Swiech, L., Kempton, H.R., Dahlman, J.E., Parnas, O., Eisenhaure, T.M., Jovanovic, M., *et al.* (2014). CRISPR-Cas9 knockin mice for genome editing and cancer modeling. Cell *159*, 440–455. http://dx.doi.org/10.1016/j.cell.2014.09.014

Pourcel, C., Salvignol, G., and Vergnaud, G. (2005). CRISPR elements in *Yersinia pestis* acquire new repeats by preferential uptake of bacteriophage DNA, and provide additional tools for evolutionary studies. Microbiology *151*, 653–663. http://dx.doi.org/10.1099/mic.0.27437-0

Prinz, F., Schlange, T., and Asadullah, K. (2011). Believe it or not: how much can we rely on published data on potential drug targets? Nat. Rev. Drug Discovery *10*, 712. http://dx.doi.org/10.1038/nrd3439-c1

Ronda, C., Maury, J., Jakočiunas, T., Jacobsen, S.A., Germann, S.M., Harrison, S.J., Borodina, I., Keasling, J.D., Jensen, M.K., and Nielsen, A.T. (2015). CrEdit: CRISPR mediated multi-loci gene integration in *Saccharomyces cerevisiae*. Microb. Cell Fact. *14*, 97. http://dx.doi.org/10.1186/s12934-015-0288-3

Sachse, C., Krausz, E., Krönke, A., Hannus, M., Walsh, A., Grabner, A., Ovcharenko, D., Dorris, D., Trudel, C., Sönnichsen, et al. (2005). High-throughput RNA interference strategies for target discovery and validation by using synthetic short interfering RNAs: functional genomics investigations of biological pathways. Methods Enzymol. *392*, 242–277. http://dx.doi.org/10.1016/S0076–6879(04)92015–0

Sander, J.D., and Joung, J.K. (2014). CRISPR-Cas systems for editing, regulating and targeting genomes. Nat. Biotechnol. *32*, 347–355. http://dx.doi.org/10.1038/nbt.2842

Sanjana, N.E., Shalem, O., and Zhang, F. (2014). Improved vectors and genome-wide libraries for CRISPR screening. Nat. Methods *11*, 783–784. http://dx.doi.org/10.1038/nmeth.3047

Schwank, G., Koo, B.K., Sasselli, V., Dekkers, J.F., Heo, I., Demircan, T., Sasaki, N., Boymans, S., Cuppen, E., van der Ent, C.K., *et al.* (2013). Functional repair of CFTR by CRISPR/Cas9 in intestinal stem cell organoids of cystic fibrosis patients. Cell. Stem. Cell. *13*, 653–658. http://dx.doi.org/10.1016/j.stem.2013.11.002

Schwartz, C.M., Hussain, M.S., Blenner, M., and Wheeldon, I. (2016). Synthetic RNA Polymerase III Promoters Facilitate High-Efficiency CRISPR-Cas9-Mediated Genome Editing in *Yarrowia lipolytica*. ACS Synth. Biol. *5*, 356–359. http://dx.doi.org/10.1021/acssynbio.5b00162

Shahana, S., Childers, D.S., Ballou, E.R., Bohovych, I., Odds, F.C., Gow, N.A., and Brown, A.J. (2014). New Clox Systems for rapid and efficient gene disruption in *Candida albicans*. PLOS One 9, e100390. http://dx.doi.org/10.1371/journal.pone.0100390

Shalem, O., Sanjana, N.E., Hartenian, E., Shi, X., Scott, D.A., Mikkelsen, T.S., Heckl, D., Ebert, B.L., Root, D.E., Doench, J.G., *et al.* (2014). Genome-scale CRISPR-Cas9 knockout screening in human cells. Science *343*, 84–87. http://dx.doi.org/10.1126/science.1247005

Shan, Q., Wang, Y., Li, J., Zhang, Y., Chen, K., Liang, Z., Zhang, K., Liu, J., Xi, J.J., Qiu, J.L., *et al.* (2013). Targeted genome modification of crop plants using a CRISPR-Cas system. Nat. Biotechnol. *31*, 686–688. http://dx.doi.org/10.1038/nbt.2650

Stovicek, V., Borodina, I., and Forster, J. (2015). CRISPR–Cas system enables fast and simple genome editing of industrial *Saccharomyces cerevisiae* strains. Metab. Eng. Commu..n. *2*, 13–22. http://dx.doi.org/10.1016/j.meteno.2015.03.001

Takahashi, T., Jin, F.J., Sunagawa, M., Machida, M., and Koyama, Y. (2008). Generation of large chromosomal deletions in koji molds *Aspergillus oryzae* and *Aspergillus sojae* via a loop-out recombination. Appl. Environ. Microbiol. *74*, 7684–7693. http://dx.doi.org/10.1128/AEM.00692-08

Tan, W., Carlson, D.F., Lancto, C.A., Garbe, J.R., Webster, D.A., Hackett, P.B., and Fahrenkrug, S.C. (2013). Efficient nonmeiotic allele introgression in livestock using custom endonucleases. Proc. Natl. Acad. Sci. U.S.A. *110*, 16526–16531. http://dx.doi.org/10.1073/pnas.1310478110

Tong, Y., Charusanti, P., Zhang, L., Weber, T., and Lee, S.Y. (2015). CRISPR-Cas9 Based Engineering of Actinomycetal Genomes. ACS Synth. Biol. *4*, 1020–1029. http://dx.doi.org/10.1021/acssynbio.5b00038

Upadhyay, S.K., Kumar, J., Alok, A., and Tuli, R. (2013). RNA-guided genome editing for target gene mutations in wheat. G3 *3*, 2233–2238. http://dx.doi.org/10.1534/g3.113.008847

van den Bosch, M., Lohman, P.H., and Pastink, A. (2002). DNA double-strand break repair by homologous recombination. Biol. Chem. *383*, 873–892. http://dx.doi.org/10.1515/BC.2002.095

Vyas, V.K., Barrasa, M.I., and Fink, G.R. (2015). A *Candida albicans* CRISPR system permits genetic engineering of essential genes and gene families. Sci. Adv. *1*, e1500248. http://dx.doi.org/10.1126/sciadv.1500248

Wang, H., Yang, H., Shivalila, C.S., Dawlaty, M.M., Cheng, A.W., Zhang, F., and Jaenisch, R. (2013). One-step generation of mice carrying mutations in multiple genes by CRISPR/Cas-mediated genome engineering. Cell *153*, 910–918. http://dx.doi.org/10.1016/j.cell.2013.04.025

Wang, J., and Quake, S.R. (2014). RNA-guided endonuclease provides a therapeutic strategy to cure latent herpesviridae infection. Proc. Natl. Acad. Sci. U.S.A. *111*, 13157–13162. http://dx.doi.org/10.1073/pnas.1410785111

Wang, T., Wei, J.J., Sabatini, D.M., and Lander, E.S. (2014). Genetic screens in human cells using the CRISPR-Cas9 system. Science *343*, 80–84. http://dx.doi.org/10.1126/science.1246981

Wang, Y., Cheng, X., Shan, Q., Zhang, Y., Liu, J., Gao, C., and Qiu, J.-L. (2014). Simultaneous editing of three homoeoalleles in hexaploid bread wheat confers heritable resistance to powdery mildew. Nat. Biotechnol. *32*, 947–951. http://dx.doi.org/10.1038/nbt.2969

Webber, P. (2014). Does CRISPR-Cas open new possibilities for patents or present a moral maze? Nat. Biotechnol. *32*, 331–333. http://dx.doi.org/10.1038/nbt.2843

Xie, K., and Yang, Y. (2013). RNA-guided genome editing in plants using a CRISPR-Cas system. Mol. Plant. *6*, 1975–1983. http://dx.doi.org/10.1093/mp/sst119

Xue, W., Chen, S., Yin, H., Tammela, T., Papagiannakopoulos, T., Joshi, N.S., Cai, W., Yang, G., Bronson, R., Crowley, D.G., *et al.* (2014). CRISPR-mediated direct mutation of cancer genes in the mouse liver. Nature *514*, 380–384. http://dx.doi.org/10.1038/nature13589

Yang, H., Wang, H., Shivalila, C.S., Cheng, A.W., Shi, L., and Jaenisch, R. (2013). One-step generation of mice carrying reporter and conditional alleles by CRISPR/Cas-mediated genome engineering. Cell *154*, 1370–1379. http://dx.doi.org/10.1016/j.cell.2013.08.022

Yosef, I., Manor, M., Kiro, R., and Qimron, U. (2015). Temperate and lytic bacteriophages programmed to sensitize and kill antibiotic-resistant bacteria. Proc. Natl. Acad. Sci. U.S.A. *112*, 7267–7272. http://dx.doi.org/10.1073/pnas.1500107112

Zhang, C., Meng, X., Wei, X., and Lu, L. (2016). Highly efficient CRISPR mutagenesis by microhomology-mediated end joining in *Aspergillus fumigatus*. Fungal. Genet. Biol. *86*, 47–57. http://dx.doi.org/10.1016/j.fgb.2015.12.007

Zhen, S., Hua, L., Liu, Y.-H., Gao, L.-C., Fu, J., Wan, D.-Y., Dong, L.-H., Song, H.-F., and Gao, X. (2015). Harnessing the clustered regularly interspaced short palindromic repeat (CRISPR)/CRISPR-associated Cas9 system to disrupt the hepatitis B virus. Gene Therapy. 22, 404–412 http://dx.doi.org/10.1038/gt.2015.2

Zhou, Y., Zhu, S., Cai, C., Yuan, P., Li, C., Huang, Y., and Wei, W. (2014). High-throughput screening of a CRISPR/Cas9 library for functional genomics in human cells. Nature *509*, 487–491. http://dx.doi.org/10.1038/nature13166

Applications of CRISPR/Cas9 in Reproductive Biology

8

Faheem Ahmed Khan, Nuruliarizki Shinta Pandupuspitasari, Huang ChunJie, Hafiz Ishfaq Ahmad, Kai Wang, Muhammad Jamil Ahmad and ShuJun Zhang*

Key Laboratory of Agricultural Animal Genetics, Breeding and Reproduction of the Ministry of Education, College of Animal Science and Technology, Huazhong Agricultural University, Wuhan, The People's Republic of China.

*Correspondence: sjxiaozhang@mail.hzau.edu.cn

https://doi.org/10.21775/9781910190630.08

Abstract

Genome editing is unravelling its benefits in wide areas of scientific development and understanding. The advances of genome editing from ZFNs and TALLENs to CRISPRs defines its wide applicability. Reproduction is the fundamental process by which all organisms maintain their generations. CRISPR/Cas9, a new versatile genome editing tool has been recently tamed to correct several disease causing genetic mutations, spreading its arms to improve reproductive health. It not only edits harmful genetic mutations but is also applied to control the spread of parasitic diseases such as malaria by introducing selfish genetic elements, propagated through generations and population via reproduction. These applications led us to review the recent developments of CRISPRs use in reproductive biology.

Introduction

The blueprint of life and heredity is written in genome. The human body consists of trillions of cells which are divided in autosomal cells and gametes (Soh *et al.*, 2014). The autosomes consists of two genomes while gametes comprises a single genome. During the process of fertilization the two gametes, one from each partner, unite to form a zygote (Hughes and Page, 2015). The genome is built up of billions of DNA base pairs subdivided into genes that interact with environment to develop a specific phenotype. Although extensive studies have been carried out but still the understanding of genes and its function is far from understanding because genes are differentially expressed. The present advances in genome editing technologies takes a leap towards endogenous genome modification, which gives rise to several opportunities to deal with diseases and undesired phenotypes (Khan *et al.*, 2016).

The continuous update to knowledge reservoir owes to molecular studies unfolding several important aspects of genes and their interactions with the environment enabling scientists to do forward and reverse genetics model cells and organisms to uncover the intricate mechanisms that may be taking place. Such studies also open avenues of its practical applications, especially in medicine. The genome manipulation was first made possible in the 1970s when *E. coli* was first manipulated for therapeutic protein production (Itakura *et al.*, 1977). The advent of gene silencing methods further increased the precise understanding of gene function that rely on homologous template. It is very precise but its application on a large scale remains inefficient. To overcome these limitations ZFNs and TALENs are developed which are proteins with DNA-binding domains (Dhanasekaran *et al.*, 2006; Reyon *et al.*, 2012) that are much more efficient than the gene silencing technologies; however, its wider application is limited by unavailability of skilled technicians who can make a special targeting protein for each genomic target.

Clustered regularly interspaced short palindromic repeats (CRISPR/Cas) is the most recent and accurate tool to target nucleic acids of any nature to modify, edit or disrupt any gene. The CRISPR/Cas system can virtually be reprogrammed against any genome of interest to achieve required goals. This special immune response is developed by a unique set of CRISPR array which contains a protospacer from invading nucleic acid recognized by short protospacer adjacent motif (PAM) which helps the host resistant to its own immune response (Bolotin *et al.*, 2005; Barrangou *et al.*, 2007). The simple machinery of CRISPR/Cas system makes it a technology of choice in various biotechnological platforms including gene therapy, genome editing and therapeutic production. The CRISPR/Cas9 provides an efficient way to make targeted double-stranded breaks (DSBs) in DNA that can knock out any gene through non-homologous DNA repair pathway (NHEJ) and can stimulate template-based homologous repair through homology directed repair (HDR) (Khan *et al.*, 2016). The CRISPR/Cas9 system requires only the design of the new sgRNA against target, as compared to ZFNs and TALENs that needs special protein for each target.

Gene drives is a technique that can enhance the inheritance of a particular gene to increase its prevalence in the population (Windbichler *et al.*, 2011). In natural populations, genes sometimes gain a fitness bias that does not depend on the organism. The genes in a sexually reproducing organisms have a 50% chance of being inherited by offspring, but some genes somehow gain an evolutionary advantage such that they are passed on to more than 50% of offspring. This phenomenon is considered as natural 'gene drives' that initially is found in single organism and then is slowly transferred, over generations, to the whole population. Artificial gene drives can be a very useful way to control the vectors of several diseases, including malaria (Hammond *et al.*, 2015), dengue and zika transmitted by mosquitoes. The CRISPR/Cas9 system can be efficiently utilized to create artificial gene drives that can render mosquitoes sterile. Apart from disease control, CRISPR/Cas9 can be used to study several fertility-related genes in males, hence providing an important platform for molecular genetic studies of fertility, development and control of disease-causing vectors. This review will focus on applications of the most versatile CRISPR/Cas9 in reproduction that is the fundamental process of species survival.

Gene drives and CRISPR/Cas9

The selfish genetic elements constitute gene drive systems that are biased towards their own inheritance in super-Mendelian fashion (Noble *et al.*, 2016; Champer *et al.*, 2016). These gene drives are considered as means to eradicate insect-borne diseases including malaria, zika and dengue (Hammond *et al.*, 2015). Naturally found gene drives are in the form of transposons, commonly known as jumping genes (Charlesworth *et al.*, 1989) or those causing segregation distortion (Tao *et al.*, 2001) and Medea elements (Chen *et al.*, 2007). The discovery of CRISPR/Cas9 genome editing technology has made it possible to create synthetic gene drives that can change the dynamics of natural populations (Champer *et al.*, 2016).

It is noteworthy that synthetic gene drives are widely studied and have potential to widely eradicate the protozoal, viral and helminthic diseases that causes a huge human disease burden (Zamanian and Andersen 2016). Synthetic gene drives to control vector-borne diseases have two principal objectives: population modification or population suppression (Burt, 2003, 2014). The eradication of the vector, for example mosquitos carrying the malaria parasite, is an example of population suppression. In contrast, if an allele important to disease resistance is introduced into a vector, or if an allele is transferred to a vector that makes it unable to transfer pathogen to host, this is a form of population modification gene drives. The concept of such synthetic gene drives is derived from natural gene drives but its progress was very slow until the use of the versatile CRISPR/Cas9 system (Zamanian and Andersen, 2016). The recent applications of CRISPR/Cas9 in *Drosophila* for mutagenic chain reaction (Gantz and Bier, 2015) and to achieve gene drives in malarial mosquito vectors that converts autocatalytic heterozygous locus to homozygous locus rendering mosquito unable to transmit malaria by spreading antimalarial effector genes through mutagenic chain reaction (Gantz *et al.*, 2015). Furthermore, this mutagenic chain reaction was utilized recently against *Plasmodium falciparum* vector *Anopheles gambiae* to spread female sterility genes, hence targeting the reproduction capability of *Anopheles gambiae* (Hammond *et al.*, 2016). These studies open powerful avenues to control several other mosquito borne vector diseases including filarial nematodes by eradicating LF genes in wild populations that can be done by creating population suppression drives by spreading genes that compromise reproductive capacity and limit the mosquito populations transmitting *Wuchereria* and *Brugia* nematodes, the main aetiological agents of LF transfer in humans (Zamanian and Andersen, 2016).

The other kind of gene drive is sex-linked meiotic drive that occurs when certain alleles are biased at meiotic level and hence are transferred to gametes and ultimately to offspring. Several such kinds of meiotic gene drives are found in nature of which sex-linked genetic drives functions by preventing the gametes lacking meiotic drive to mature and help skew gender ratios (Helleu et *al.*, 2014; Champer *et al.*, 2016). The mechanisms by which the drives work is not fully understood, but mostly an X-shredder mechanism is widely accepted. If the X-shredder gene is on the Y chromosome it will quickly spread in a population and will suppress the population. Often an endonuclease targeting X chromosome is designed for such drives in heterogametic X/Y species. A schematic illustrating endonuclease gene drives is shown in Fig. 8.1.

These approaches emphasizes the role of CRISPR/Cas9 genome editing technology beyond the study of fertility related genes and mechanistic pathways but provide a robust platform to spread desired gene drives efficiently in natural populations to control various vector-borne diseases of human health importance.

Genome engineering and reproductive biology

The gene modification in primates remains a challenge because of their long life cycles and hence make editing impractical. The gene introduction in primates is performed mostly with viral vectors that resulted in transgenic monkeys, however, the site of gene insertion and its number of copies are not controlled (Schubert, 2014). These limitations are overdone by CRISPR/Cas9 when Sha and colleagues established a technique where CRISPR system target specific DNA sequences and produce single-stranded DNA cuts at RNA-specified targets (Schubert, 2014). Three genes were targeted

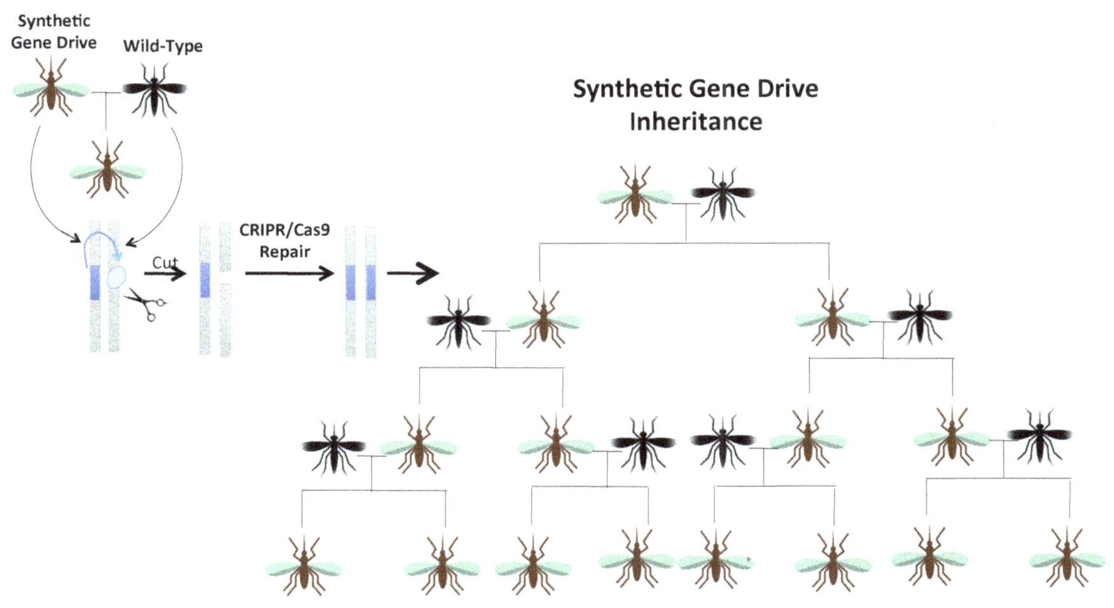

Figure 8.1 A schematic representing endonuclease gene drive using CRISPR/Cas9 system. A strategy to fight vector-borne diseases. Modified from Dicarlo *et al.*, 2015.

in monkey at one cell stage embryo in which two genes, *PRAG* and *RAG1*, were shown successfully knocked out in twins simultaneously (Niu *et al.*, 2014).

The recent years have witnessed gene editing in human embryos as the CRISPR/Cas9 come up with its applications in reproductive medicine with proof-of-principle to generate tripronuclear (3PN) zygotes (Liang *et al.*, 2015a). The results of Liang and colleagues show the use of CRISPR/Cas9 to efficiently cleave β-globin gene (*HBB*), although the homology directed repair (HDR) was found to be low, T7E1 assay and whole exome sequencing reveals its off-target effects that needs to be improved in future studies for its clinical applications. The reports that show correction of disease causing mutations for tyrosinaemia (Yin *et al.*, 2014), Duchenne muscular dystrophy (Khan *et al.*, 2016) by CRISPR systems open gates for its application to diseases causing infertility both in male and females.

A possible alternative to zygote editing is the editing of male and female germ cells at an immature stage by applying CRISPR/Cas9 system to get corrected mature sperms and oocytes which can be used in artificial assisted reproduction (Vassena *et al.*, 2016). The recent advances has made the genome editing more feasible in germlines, embryos, gametes and gamete precursors. To deal with the monogenic diseases, the mutations of which are often well characterized, and if can be corrected at sperm or at oocyte level, will produce disease-free individuals.

The 50% of genes related to fertility will not produce phenotype of fertility or infertility; hence it is a long tedious job to screen those phenotypes through generations. The fast-track CRISPR/Cas9 system can accomplish this task in months to determine whether a gene needs further study or not with its ability to precisely disrupt or induce multiple mutations to characterize genes, although redundancy of reproductive genes remains a challenge which can be overcome by properly designing the CRISPR/Cas9 system (Young *et al.*, 2015). Similarly, the effect of multiple genes on reproduction can be determined by utilizing the CRISPR/Cas9 system, as in the case of growth differentiation factor (Gdf9) and bone morphogenetic protein (Bmp15), two unrelated genes (Yan *et al.*, 2001). CRISPR/Cas9 system will help develop knock-out models of mice and other animals for several genes as well as combined functioning of genes in pathways and its epigenetic interactions.

CRISPR/Cas system in male reproduction

The gene manipulation research in animal models has played an important role in assessment of reproduction processes. Reproduction is one of the best suited biological systems to which gene manipulation such as knock-out can be successfully applied. The spermatogenic or haploid germ cell genes that are important for fertility are more specific to the gonads and often comprise a single exon (Ikawa *et al.*, 2010). The breeding system involved in creating homozygous genetic manipulations in genes involved in gametogenesis, fertilization and conception. Whatever the reason, there are highly specific reproductively related genes of which the function is unveiled by making the use of gene disruption techniques to explore the vital factors of fertility. Gene disruption methods have long been advantageous to study male reproductive biology. However, because of the time and cost involved, this technology was not a practicable technique except in proficient laboratories. Genome editing tools such as CRISPR/Cas9 system has been extensively used to modify genes in organisms including human and animal cells, and hold a remarkable potential for clinical research applications. To date, an information gap remains in understanding of DNA repair mechanism in human embryos and in the efficiency and potential off-target effects using CRISPR/Cas9 technologies in human pre-implantation zygotes and early embryos (Liang *et al.*, 2015a). The introduction of the CRISPR/Cas9 system for gene disruption, has facilitated the modern era of genetic investigation. The application of CRISPR/Cas9 technology to generate gene disrupted mouse models to investigate the genetics of male fertility that look at the essential genetic component of male reproductive system (Young *et al.*, 2015). An important aspect to consider in the utilization of CRISPR/Cas9 system in reproduction is that 50% of fertility genes produce no distinct phenotype when disrupted (Okabe, 2014), fears no conclusion. With ease and precision of CRISPR/Cas9 system this has revolved around, with worth perusing ability of producing fertility or infertility phenotype within few months, in addition to introduction of point mutation, ability to further investigate the molecular and epigenetic mechanisms of fertility related genes (Mashiko *et al.*, 2013). Currently the CRISPR/Cas9 systems are used to generate knockout mice to study the role of individual or multiple genes in reproduction where non-homologous end joining repair mechanism (NHEJ) is activated by DNA cleavage. The CRISPR/Cas9 systems are also used to introduce point mutations in mouse models of human conditions via homologous directed repair (HDR) mechanisms, as well as being used to analyse complex aspects of post-translational modifications (Young *et al.*, 2015). The use of CRISPR/Cas9 system circumvented these issue by allowing scientists to create multiple site mutations simultaneously which was previously impossible via conventional methods (Archambeault and Matzuk, 2014; Capecchi, 2005).

With the development of CRISPR/Cas9 system the knockout mouse models of testis specific genes are generated to study spermatogenesis process in male reproductive systems *in vivo*. The function of testis-specific component, *Slx2* in spermatogenesis was investigated by disrupting the *Slx2* gene with CRISPR/Cas9 system and knockout male mouse was obtained and is compared with the testis-specific genes of the wild type male mouse. Taken together, it was revealed that *slx2* itself does not play an essential role in spermatogenesis and the CRISPR/Cas9 technology can boost up the functional study of X-linked genes of testis *in vivo* (Li *et al.*, 2015).

CRISPR/Cas9 system could be used on growing sperm cells to produce gene-corrected mature sperm cells, which could be used for assisted reproduction. In male germline cells, it is not possible that mature post-meiotic sperms could be made subject to genetic manipulations, however, spermatogonial stem cells seem a better option for manipulation, particularly for the patients suffering from infertility having no sperm cells due to maturation arrest. Current progresses in *in vitro* spermatogonia stem cells culture seem to bring that prospect closer (Nickkholgh *et al.*, 2014). In a study, the *Crygc* gene in spermatogonial stem cells are mutated via CRISPR/Cas9 system. The mutated spermatogonial stem cells undergone spermatogenesis after transplantation into seminiferous tubules of infertile mouse testes. Moreover, the disease causing mutations in *Crygc* gene in spermatogonial stem cells is repaired by non-homologous end joining or homology end repair mechanisms induced by CRISPR/Cas9 in mouse, resulting in spermatogonial stem cells carrying modified gene with no off-target mutations at entire genome sequencing and the offspring showed 100% efficient phenotype (Wu *et al.*, 2015). A better substitute for genetically modified spermatogonial stem cells seems to be the unprecedentedly simple CRISPR/Cas9 system (Chapman *et al.*, 2015; Sato *et al.*, 2015). Furthermore, the CRISPR/Cas9 system not only circumvents the nuclease engineering but also generates fewer off-target effects than ZFNs (Ul Ain *et al.*, 2015). According to recent studies, no obvious off-target genetic effects could be identified in SSCT experiments involving CRISPR/Cas9-mediated gene targeting transplantation of genetically modified SSC (Wu *et al.*, 2015).

Interestingly, CRISPR/Cas9 system has been shown to be successful in repairing the disease causing mutations in various species such as the *Crygc* gene mutation in mice which causes cataracts (Wu *et al.*, 2013), dystrophin gene which causes Duchenne muscular dystrophy (*DMD*) (Long *et al.*, 2014) and *Fah* mutation in hepatocytes are repaired by CRISPR/Cas9 system. As well as in humans, the CRISPR/Cas9 system has been used to correct β-thalassaemia, haemoglobin-β and dystrophin gene in pluripotent stem cells (Xie *et al.*, 2014). CRISPR/Cas9 catalysed the rat spermatogonial stem cells to regenerate spermatogenesis and showed long-term sperm-producing capability following transplantation in rat testes. Targeted mutations in germ line cells such as epithelial stromal interaction 1 (*Epsti1*) and receptor tyrosine protein kinase 3 (*Erbb3*) were vertically transmitted from recipient to exclusive produce non-mosaic offspring. Monoclonal development of *Erbb3*null germlines exposed recessive spermatogenesis defects and yielded modified offspring isogenic at targeted alleles. Therefore, spermatogonial gene modification with CRISPR/Cas9 provided a platform to produce targeted germ line modifications in rats, and to study spermatogenesis (Chapman *et al.*, 2015).

CRISPR/Cas system in female reproduction

CRISPR/Cas9 technology is especially useful in developmental studies in model organisms and the efficiency and specificity of CRISPR/Cas9 system mediated the genome editing in human embryonic cells. Recently, a group used CRISPR/Cas9 in human tripronuclear zygotes with one oocyte nucleus and two sperm nuclei to cleave the endogenous β-globin gene and repaired the β-globin gene through the non-crossover HDR mechanism (Liang *et al.*, 2015a). The embryos are edited directly through cytoplasmic pronuclei microinjection of zygote with no off-target genetic modifications. Although the efficiency of genome editing in embryos is low, but still several studies have been done in different animals such as rat, sheep, cattle, pigs and dogs to demonstrate the feasibility and efficiency of genome editing in animals (Shao *et al.*, 2014; Zou *et al.*, 2015; Jeong *et al.*, 2016). This technique can prevent the chances of genetic disorders, for example a study on cataract development demonstrated successful 1 bp deletion in a specific gene in mouse offspring (Wu *et al.*, 2013).

Microinjection of Cas9 or TALENs in non-human primates led to the birth of modified offspring (Niu *et al.*, 2014). The efficiency of genetic modifications by Cas9 or TALENs in mammalian zygote ranges from 0.5% to 40.9% per injected zygote (Araki and Ishii, 2014). Recently, CRISPR/Cas9 editing technique was performed in human zygote to validate its specificity and fidelity (Liang *et al.*, 2015b). This group injected 86 3PN zygotes along with other DNA molecules designated as new DNA with CRISPR/Cas9 and 28 genome-edited zygotes were successfully spliced, and that only four (5.6% of the total) contained the correct genetic material inserted through homologous recombination. The edited embryos were mosaic, with results similar to findings in other model systems (Yen *et al.*, 2014). Furthermore, a few number of off-target mutations were found, which were supposed to have been introduced by CRISPR/Cas9 complex acting in another part of genome. Notably, a specific portion of genome known as exome was verified for off-target mutations and such mutations could be higher due to mosaic nature of gene correction in edited embryos (Vassena *et al.*, 2016). As a possible alternative to the zygote approach, genome modifications could also be applied during gametogenesis. In this aspect, the CRISPR/Cas9 system could be applied on growing oocytes to generate genetically modified oocytes, which would be subsequently useful for assisted reproductive technology, thus the off-target mutations could be overcome in the following generation (Vassena *et al.*, 2016). In female germ line cells, the oocyte is easily accessible for genetic modifications. However, *in vitro* maturation of oocytes at germinal vesicle stage is applied for necessary manipulation, though efficiency and accuracy of CRISPR/Cas9 should be investigated during meiosis (Vassena *et al.*, 2016). The schematic illustrating the CRISPR/Cas9 editing is shown in Fig. 8.2. The mutations leading to monogenic diseases such as cystic fibrosis are well characterized in patients and these mutations can be corrected in germ line cells and patients can produce oocytes which are free from mutations. Therefore, they produce healthy embryos and offspring that are not carrier for these disease causing mutations in the population. (Vassena *et al.*, 2016). Besides this, recently a group used CRISPR/Cas9 technology to modify the pig genome and produced genetically modified pigs by introducing CRISPR/Cas9 mutated CD1D or CD163 into somatic cells. The CRISPR/Cas9 was introduced *in vitro* produced porcine zygote and the system was effective in generating mutations in CD1D, CD163 and eGFP with 100% targeting accuracy in embryos at blastocyst stage. Direct inoculation of CRISPR/Cas9 targeting CD1D or CD163 into zygotes resulted in offspring which have mutations on both alleles with only one CD1D pig having a mosaic genotype. It was concluded that the CRISPR/Cas9 system can be used by two methods. The CRISPR/Cas9 can be used to modify somatic cells followed by somatic cell nuclear transfer as well as the CRISPR/Cas9 components can also be used in *in vitro*-produced zygotes to generate pigs with specific genetic modifications (Whitworth *et al.*, 2014).

In addition to its application in genome editing, the feasibility and specificity of CRISPR/Cas9 technology offers unique opportunities to expedite the control of insect vector diseases through the development of gene drive system (Esvelt *et al.*, 2014). A CRISPR/Cas9 based translation for the control of human malaria insect vector involves the development of endonuclease based gene drive

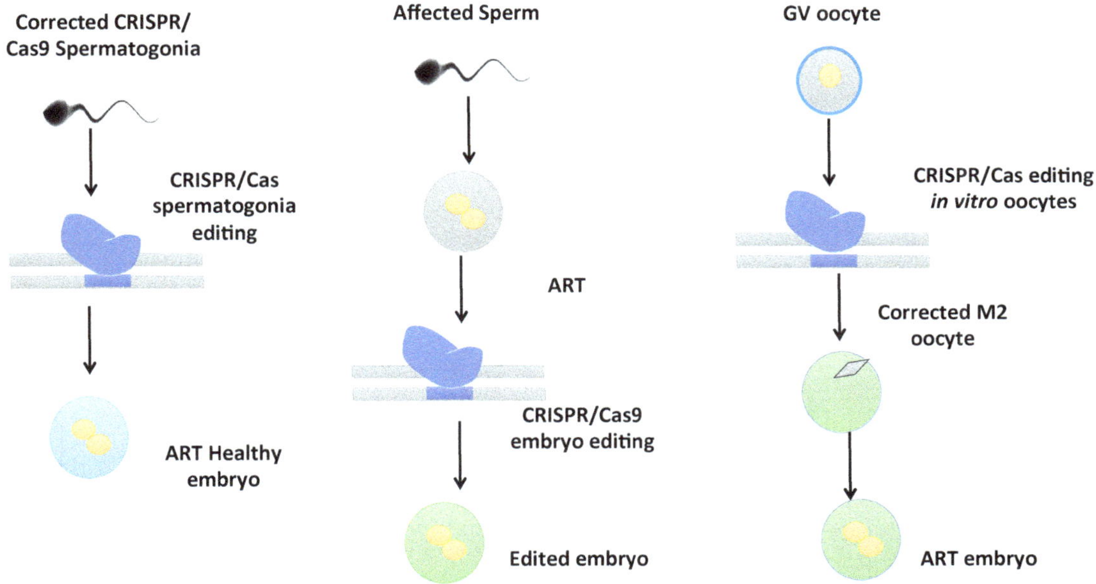

Figure 8.2 A schematic illustrating the use of CRISPR/Cas9 in sperm, oocyte and embryo genome editing. The corrected sperms and oocytes can be used in artificial reproduction techniques to produce healthy embryos.

systems that interfere with disease transmitting ability of *Anopheles gambiae* mosquitoes, that could be attained by blocking or reducing the reproductive ability of parasite or insect vector. A CRISPR/Cas based gene drive system was developed in both male and female sexes of human malaria vector *A. gambiae* to yield recessive mutation in female fertility genes before or during the gamete formation process (Deredec *et al.*, 2008; Hammond *et al.*, 2016).

CRISPR/Cas technology has opened up opportunities for the correction of human mitochondrial diseases by editing mitochondrial DNA in mouse oocyte incorporated into patient cells. This riveting progress may inspire the development of unique remedy for human maternally inherited diseases. The highly efficient CRISPR/Cas-targeted mutant mitochondrial DNA in both human and animal model cells revealed in future healthy babies born from oocytes of patients where mutant mtDNA is cleaned and the residues of mutant DNA is reduced below the level required for disease manifestation. Regardless of all potential benefits narrated, the authors also instructed a risk that embryos might be unable to implant in uterus when mtDNA copy number in the 'amended' embryos was below a specific threshold (Wang *et al.*, 2015).

In comparison with former techniques, current genome editing technologies particularly CRISPR/Cas9 is not only precise but also inexpensive and very efficient. Further, there is increased acceptance and use of techniques for assisted reproduction, which are likely to be required for genome editing in human embryos.

Conclusion

The CRISPR/Cas system is revolutionizing the speed of genome editing of disease-causing mutations and is also capable of producing corrected mature sperms and oocytes that can be used to generate disease-free individuals. It has also been efficiently applied to produce synthetic gene drives in populations to control earlier and emerging vector-borne diseases. The safe applications of CRISPRs in reproduction can guarantee huge reduction in genetic as well as vector-borne parasitic diseases.

Acknowledgements

Financial assistance from the natural Science Foundation of China (3157236, 31272427) are greatly appreciated. the constructive discussions of laboratory colleagues are highly appreciated.

References

Araki, M. & Ishii, T. 2014. International regulatory landscape and integration of corrective genome editing into in vitro fertilization. Reproductive Biol. Endocrinol., *12*, 1.

Archambeault, D.R., and Matzuk, M.M. (2014). Disrupting the male germ line to find infertility and contraception targets. Ann. Endocrinol. *75*, 101–108. http://dx.doi.org/10.1016/j.ando.2014.04.006

Barrangou, R., Fremaux, C., Deveau, H., Richards, M., Boyaval, P., Moineau, S., Romero, D.A., and Horvath, P. (2007). CRISPR provides acquired resistance against viruses in prokaryotes. Science, *315*, 1709–1712.

Bolotin, A., Quinquis, B., Sorokin, A., and Ehrlich, S.D. (2005). Clustered regularly interspaced short palindrome repeats (CRISPRs) have spacers of extrachromosomal origin. Microbiology, *151*, 2551–2561.

Burt, A. (2003). Site-specific selfish genes as tools for the control and genetic engineering of natural populations. Proc. Biol. Sci.*270*, 921–928. http://dx.doi.org/10.1098/rspb.2002.2319

Burt, A. (2014). Heritable strategies for controlling insect vectors of disease. Phil. Trans. R. Soc. Lond. B. Biol. Sci.*369*, 20130432. http://dx.doi.org/10.1098/rstb.2013.0432

Capecchi, M.R. (2005). Gene targeting in mice: functional analysis of the mammalian genome for the twenty-first century. Nat. Rev. Genet.*6*, 507–512.

Champer, J., Buchman, A., and Akbari, O.S. (2016). Cheating evolution: engineering gene drives to manipulate the fate of wild populations. Nat. Rev. Genet. *17*, 146–159. http://dx.doi.org/10.1038/nrg.2015.34

Chapman, K.M., Medrano, G.A., Jaichander, P., Chaudhary, J., Waits, A.E., Nobrega, M.A., Hotaling, J.M., Ober, C., and Hamra, F.K. (2015). Targeted germline modifications in rats using CRISPR/Cas9 and spermatogonial stem cells. Cell. Rep.*10*, 1828–1835. http://dx.doi.org/10.1016/j.celrep.2015.02.040

Deredec, A., Burt, A., and Godfray, H.C. (2008). The population genetics of using homing endonuclease genes in vector and pest management. Genetics*179*, 2013–2026. http://dx.doi.org/10.1534/genetics.108.089037

Dhanasekaran, M., Negi, S., and Sugiura, Y. (2006). Designer zinc finger proteins: tools for creating artificial DNA-binding functional proteins. Acc. Chem. Res. 39, 45–52. http://dx.doi.org/10.1021/ar050158u

DiCarlo, J.E., Chavez, A., Dietz, S.L., Esvelt, K.M., and Church, G.M., 2015. RNA-guided gene drives can efficiently and reversibly bias inheritance in wild yeast. *bioRxiv*, 013896.

Esvelt, K. M., Smidler, A. L., Catteruccia, F. & Church, G. M. 2014. Concerning RNA-guided gene drives for the alteration of wild populations. Elife, 3, e03401.

Khan, F.A., Pandupuspitasari, N.S., Chun-Jie, H., Ao, Z., Jamal, M., Zohaib, A., Khan, F.A., Hakim, M.R., and ShuJun, Z. (2016). CRISPR/Cas9 therapeutics: a cure for cancer and other genetic diseases. Oncotarget, 7, 52541-52552. http://dx.doi.org/10.18632/oncotarget.9646

Gantz, V.M., and Bier, E. (2015). Genome editing. The mutagenic chain reaction: a method for converting heterozygous to homozygous mutations. Science *348*, 442–444. http://dx.doi.org/10.1126/science.aaa5945

Gantz, V.M., Jasinskiene, N., Tatarenkova, O., Fazekas, A., Macias, V.M., Bier, E., and James, A.A. (2015). Highly efficient Cas9-mediated gene drive for population modification of the malaria vector mosquito *Anopheles stephensi*. Proc. Natl. Acad. Sci. U.S.A. *112*, E6736–43. http://dx.doi.org/10.1073/pnas.1521077112

Hammond, A., Galizi, R., Kyrou, K., Simoni, A., Siniscalchi, C., Katsanos, D., Gribble, M., Baker, D., Marois, E. & Russell, S. 2016. A CRISPR-Cas9 gene drive system targeting female reproduction in the malaria mosquito vector *Anopheles gambiae*. Nature biotechnology, 34, 78–83.

Helleu, Q., Gérard, P.R., and Montchamp-Moreau, C. (2014). Sex chromosome drive. Cold Spring Harbor Perspect. Biol. 7, a017616. http://dx.doi.org/10.1101/cshperspect.a017616

Hughes, J.F., and Page, D.C. (2015). The biology and evolution of mammalian y chromosomes. Annu. Rev. Genet. *49*, 507–527. http://dx.doi.org/10.1146/annurev-genet-112414-055311

Ikawa, M., Inoue, N., Benham, A.M., and Okabe, M. (2010). Fertilization: a sperm's journey to and interaction with the oocyte. J. Clin. Invest. *120*, 984–994. http://dx.doi.org/10.1172/JCI41585

Itakura, K., Hirose, T., Crea, R., Riggs, A.D., Heyneker, H.L., Bolivar, F., and Boyer, H.W. (1977). Expression in *Escherichia coli* of a chemically synthesized gene for the hormone somatostatin. Science *198*, 1056–1063.

Jeong, Y.-H., Kim, Y. J., Kim, E. Y., Kim, S. E., Kim, J., Park, M. J., Lee, H.-G., Park, S. P. & Kang, M.-J. 2016. Knock-in fibroblasts and transgenic blastocysts for expression of human FGF2 in the bovine β-casein gene locus using CRISPR/Cas9 nuclease-mediated homologous recombination. Zygote *24*, 442–456.

Li, M., Huang, R., Jiang, X., Chen, Y., Zhang, Z., Zhang, X., Liang, P., Zhan, S., Cao, S., Songyang, Z., *et al.* (2015). CRISPR/Cas9 promotes functional study of testis specific X-linked gene in Vivo. PLOS One*10*, e0143148. http://dx.doi.org/10.1371/journal.pone.0143148

Liang, P., Xu, Y., Zhang, X., Ding, C., Huang, R., Zhang, Z., Lv, J., Xie, X., Chen, Y., Li, Y., *et al.* (2015a). CRISPR/Cas9-mediated gene editing in human tripronuclear zygotes. Protein. Cell. *6*, 363–372. http://dx.doi.org/10.1007/s13238-015-0153-5

Liang, X., Potter, J., Kumar, S., Zou, Y., Quintanilla, R., Sridharan, M., Carte, J., Chen, W., Roark, N., Ranganathan, S., *et al.* (2015b). Rapid and highly efficient mammalian cell engineering via Cas9 protein transfection. J. Biotechnol. *208*, 44–53. http://dx.doi.org/10.1016/j.jbiotec.2015.04.024

Long, C., McAnally, J.R., Shelton, J.M., Mireault, A.A., Bassel-Duby, R., and Olson, E.N. (2014). Prevention of muscular dystrophy in mice by CRISPR/Cas9-mediated editing of germline DNA. Science *345*, 1184–1188. http://dx.doi.org/10.1126/science.1254445

Mashiko, D., Fujihara, Y., Satouh, Y., Miyata, H., Isotani, A., and Ikawa, M. (2013). Generation of mutant mice by pronuclear injection of circular plasmid expressing Cas9 and single guided RNA. Sci. Rep. 3, 3355. http://dx.doi.org/10.1038/srep03355

Nickkholgh, B., Mizrak, S.C., van Daalen, S.K., Korver, C.M., Sadri-Ardekani, H., Repping, S., and van Pelt, A.M. (2014). Genetic and epigenetic stability of human spermatogonial stem cells during long-term culture. Fertil. Steril. *102*, 1700–1707.e1. http://dx.doi.org/10.1016/j.fertnstert.2014.08.022

Niu, Y., Shen, B., Cui, Y., Chen, Y., Wang, J., Wang, L., Kang, Y., Zhao, X., Si, W., Li, W., *et al.* (2014). Generation of gene-modified cynomolgus monkey via Cas9/RNA-mediated gene targeting in one-cell embryos. Cell *156*, 836–843. http://dx.doi.org/10.1016/j.cell.2014.01.027

Niu, J., Zhang, B., and Chen, H. (2014). Applications of TALENs and CRISPR/Cas9 in human cells and their potentials for gene therapy. Mol. Biotechnol. *56*, 681–688. http://dx.doi.org/10.1007/s12033-014-9771-z

Noble, C., Min, J., Olejarz, J., Buchthal, J., Chavez, A., Smidler, A.L., DeBenedictis, E.A., Church, G.M., Nowak, M.A., and Esvelt, K.M. (2016). Daisy-chain gene drives for the alteration of local populations. bioRxiv, p.057307.

Okabe, M. (2014). Mechanism of fertilization: a modern view. Exp. Anim. *63*, 357–365.

Reyon, D., Tsai, S.Q., Khayter, C., Foden, J.A., Sander, J.D., and Joung, J.K. (2012). FLASH assembly of TALENs for high-throughput genome editing. Nat. Biotechnol. *30*, 460–465. http://dx.doi.org/10.1038/nbt.2170

Sato, T., Sakuma, T., Yokonishi, T., Katagiri, K., Kamimura, S., Ogonuki, N., Ogura, A., Yamamoto, T., and Ogawa, T. (2015). Genome editing in mouse spermatogonial stem cell lines using TALEN and double-nicking CRISPR/Cas9. Stem Cell Rep. *5*, 75–82. http://dx.doi.org/10.1016/j.stemcr.2015.05.011

Shao, Y., Guan, Y., Wang, L., Qiu, Z., Liu, M., Chen, Y., Wu, L., Li, Y., Ma, X., Liu, M., *et al.* (2014). CRISPR/Cas-mediated genome editing in the rat via direct injection of one-cell embryos. Nat. Protoc. *9*, 2493–2512. http://dx.doi.org/10.1038/nprot.2014.171

Schubert, C., 2014. World of reproductive biology: precise gene engineering achieved in monkeys. BOR http://dx.doi.org/10.1095/biolreprod.114.118364

Soh, Y.Q., Alföldi, J., Pyntikova, T., Brown, L.G., Graves, T., Minx, P.J., Fulton, R.S., Kremitzki, C., Koutseva, N., Mueller, J.L., *et al.* (2014). Sequencing the mouse Y chromosome reveals convergent gene acquisition and amplification on both sex chromosomes. Cell *159*, 800–813. http://dx.doi.org/10.1016/j.cell.2014.09.052

Ul Ain, Q., Chung, J.Y., and Kim, Y.H. (2015). Current and future delivery systems for engineered nucleases: ZFN, TALEN and RGEN. J. Control. Release *205*, 120–127. http://dx.doi.org/10.1016/j.jconrel.2014.12.036

Vassena, R., Heindryckx, B., Peco, R., Pennings, G., Raya, A., Sermon, K., and Veiga, A. (2016). Genome engineering through CRISPR/Cas9 technology in the human germline and pluripotent stem cells. Hum. Reprod. Update *22*, 411–419. http://dx.doi.org/10.1093/humupd/dmw005

Wang, S., Yi, F., and Qu, J. (2015). Eliminate mitochondrial diseases by gene editing in germ-line cells and embryos. Protein. Cell *6*, 472–475. http://dx.doi.org/10.1007/s13238-015-0177-x

Whitworth, K.M., Lee, K., Benne, J.A., Beaton, B.P., Spate, L.D., Murphy, S.L., Samuel, M.S., Mao, J., O'Gorman, C., Walters, E.M., *et al.* (2014). Use of the CRISPR/Cas9 system to produce genetically engineered pigs from in vitro-derived oocytes and embryos. Biol. Reprod. *91*, 78. http://dx.doi.org/10.1095/biolreprod.114.121723

Windbichler, N., Menichelli, M., Papathanos, P.A., Thyme, S.B., Li, H., Ulge, U.Y., Hovde, B.T., Baker, D., Monnat, R.J., Burt, A., and Crisanti, A. (2011). A synthetic homing endonuclease-based gene drive system in the human malaria mosquito. Nature *473*, 212–215.

Wu, Y., Liang, D., Wang, Y., Bai, M., Tang, W., Bao, S., Yan, Z., Li, D., and Li, J. (2013). Correction of a genetic disease in mouse via use of CRISPR-Cas9. Cell. Stem. Cell.*13*, 659–662. http://dx.doi.org/10.1016/j.stem.2013.10.016

Wu, Y., Zhou, H., Fan, X., Zhang, Y., Zhang, M., Wang, Y., Xie, Z., Bai, M., Yin, Q., Liang, D., *et al.* (2015). Correction of a genetic disease by CRISPR-Cas9-mediated gene editing in mouse spermatogonial stem cells. Cell Res. *25*, 67–79. http://dx.doi.org/10.1038/cr.2014.160

Xie, F., Ye, L., Chang, J.C., Beyer, A.I., Wang, J., Muench, M.O., and Kan, Y.W. (2014). Seamless gene correction of β-thalassemia mutations in patient-specific iPSCs using CRISPR/Cas9 and piggyBac. Genome Res. *24*, 1526–1533. http://dx.doi.org/10.1101/gr.173427.114

Yan, C., Wang, P., DeMayo, J., DeMayo, F.J., Elvin, J.A., Carino, C., Prasad, S.V., Skinner, S.S., Dunbar, B.S., Dube, J.L., and Celeste, A.J. (2001). Synergistic roles of bone morphogenetic protein 15 and growth differentiation factor 9 in ovarian function. Mol. Endocrinol. *15*, 854–866.

Yen, S.T., Zhang, M., Deng, J.M., Usman, S.J., Smith, C.N., Parker-Thornburg, J., Swinton, P.G., Martin, J.F., and Behringer, R.R. (2014). Somatic mosaicism and allele complexity induced by CRISPR/Cas9 RNA injections in mouse zygotes. Dev. Biol. *393*, 3–9. http://dx.doi.org/10.1016/j.ydbio.2014.06.017

Yin, H., Xue, W., Chen, S., Bogorad, R.L., Benedetti, E., Grompe, M., Koteliansky, V., Sharp, P.A., Jacks, T., and Anderson, D.G. (2014). Genome editing with Cas9 in adult mice corrects a disease mutation and phenotype. Nat. Biotechnol. *32*, 551–553. http://dx.doi.org/10.1038/nbt.2884

Young, S.A., Aitken, R.J., and Ikawa, M. (2015). Advantages of using the CRISPR/Cas9 system of genome editing to investigate male reproductive mechanisms using mouse models. Asian. J. Androl. *17*, 623–627. http://dx.doi.org/10.4103/1008-682X.153851

Zamanian, M., and Andersen, E.C. (2016). Prospects and challenges of CRISPR/Cas genome editing for the study and control of neglected vector-borne nematode diseases. FEBS. J. *283*, 3204–3221. http://dx.doi.org/10.1111/febs.13781

Zou, Q., Wang, X., Liu, Y., Ouyang, Z., Long, H., Wei, S., Xin, J., Zhao, B., Lai, S., Shen, J., *et al.* (2015). Generation of gene-target dogs using CRISPR/Cas9 system. J. Mol. Cell. Biol.*7*, 580–583. http://dx.doi.org/10.1093/jmcb/mjv061

Ethical Issues Regarding CRISPR-mediated Genome Editing

9

Zabta Khan Shinwari[1,2*], Faouzia Tanveer[1] and Ali Talha Khalil[1]

[1]Department of Biotechnology, Quaid-i-Azam University, Islamabad.
[2]Pakistan Academy of Sciences, Islamabad, Pakistan.

*Correspondence: shinwari2008@gmail.com

https://doi.org/10.21775/9781910190630.09

Abstract

CRISPR/Cas9 has emerged as a simple, precise and most rapid genome editing technology. With a number of promising applications ranging from agriculture and environment to clinical therapeutics, it is greatly transforming the field of molecular biology. However, there are certain ethical, moral and safety concerns related to the attractive applications of this technique. The most contentious issues concerning human germline modifications are the challenges to human safety and morality such as risk of unforeseen, undesirable effects in clinical applications particularly to correct or prevent genetic diseases, matter of informed consent and the risk of exploitation for eugenics. Stringent regulations and guidelines as well as worldwide debate and awareness are required to ensure responsible and wise use of CRISPR mediated genome editing technology. There is a need for an extensive dialogue among scientists, ethicists, industrialists and policy makers on its societal implications. The opinion of different elements of the society including the general public as well as religious scholars is also critical. In countries with existing legislative framework, it might be appropriate to allow CRISPR-based research to proceed with proper justification. However, much anticipated future clinical applications must be strictly regulated with newly established regulations.

Introduction

The quest for introducing the site-specific changes in the DNA sequence began when DNA was first discovered. Progress in genome engineering technologies began in 1990s now reaching to a highly advanced, easy, economical and sophisticated method for editing genomes called CRISPR/Cas9. CRISPR (Clustered Regularly Interspaced Palindromic Repeats) technology does not arrive as a breakthrough technology for editing the genomes but other genome editing platforms like TALENS (Transcription Activator-Like Effector Nucleases) and ZFN (Zinc Finger Nucleases) were in use for some time but have lost their popularity because of their complexity, expensiveness and time consumption (Carroll and Charo, 2015; Doudna and Charpentier, 2014; Hsu *et al.*, 2014; Jinek *et al.*, 2012). The aforementioned techniques require protein engineering while CRISPR needs a guide RNA molecule that recognize the target sequence (Caplan *et al.*, 2015). The genome engineering tools offer promising advances in medicine, healthcare, agriculture and food. CRISPR/Cas9 has brought a revolution in the biological innovations. This technology can be used to treat the genetically inherited diseases by correcting the responsible mutation, analysing cancer progression and genetic rearrangements (Doudna and Charpentier, 2014). Despite the wide array of applications, ethicist raises serious reservations about the CRISPR ability to easily modify the human germline cells, making these changes transmittable to the progeny.

CRISPR was initially discovered in the genome of *E. coli* by Japanese researchers in 1987 and described as short direct repeats interspaced with short sequences. Later on CRISPR was discovered in many bacteria and archaea with a possible role in gene regulation or DNA repair (Guy *et al.*, 2004; Ishino *et al.*, 1987; Makarova *et al.*, 2002; Mojica *et al.*, 2000). In 2005, it was discovered that many spacer sequences that are accompanied with CRISPRs were the derivatives of plasmid and viral sequences (Bolotin *et al.*, 2005). It was proposed by various research groups that CRISPR/Cas9 has a possible role in adaptive immunity and this proposition was confirmed later on in 2007 by researchers working on *Streptococcus thermophilus*. In 2008, Brouns *et al.*, in a study published in *Science*, showed that Cas proteins interfere with the multiplication of a virus inside prokaryotes (*E. coli*), while in the same year the DNA-targeting ability of CRISPR/Cas9 was established (Brouns *et al.*, 2008; Marraffini and Sontheimer, 2008). (For a comprehensive history and mechanism which is behind the scope of this review we recommend 'The new frontier of genome engineering with CRISPR/Cas9' by Doudna *et al.*, 2014.)

Ethical issues regarding CRISPR technology

The rapidly growing technology of genome editing, CRISPR/Cas9, is transforming the field of molecular biology enabling scientists to make desired changes in DNA in a variety of organisms. Soon after it was introduced in 2012, it has been quickly adopted due to ease of its use and simplicity. It is being considered for a variety of applications from agriculture to clinical therapeutics including human germline alterations to correct genetic diseases (Carroll and Charo, 2015). Although the ethical debate on human genetic modifications is not new (as mentioned previously), however, CRISPR/Cas9-mediated genome editing has given it a new edge. Owing to unpredictable and far reaching consequences associated with attractive applications of this technology, a comprehensive dialogue is needed on its ethical and societal implications. The present review discusses the ethical concerns related to this new technology with special emphasis on its possibilities and concerns on human germline modifications. It urges the need for an extensive dialogue among scientists, ethicists, industrialists and policy makers. The opinion of different elements of the society including the general public as well as religious scholars is critical.

Recent ethical debate: from where it started

The rapidly expanding scope of applications of CRISPR/Cas9 is incredible. It has provided a number of possibilities in molecular biology research, e.g. to turn off or on genes for studying their function or to induce mutations in cells to study how and why cells become cancerous. It can be used to alter genes in plants and animals to create drought resistant crops or make police dogs become more muscular. Then comes a rather contentious application to permanently modify the human genome to eliminate the disease-causing mutations or even enhance or introduce desired characteristics in children by adding useful genes. Though the idea of genetically modified humans and designer babies has been popular for a long time in both science and fiction, CRISPR provided a reasonable tool to turn this long cherished dream into a reality (le Page, 2015). Genomic changes made in the non-reproductive cells are not heritable but if alterations are done to the germ cells (that develop into eggs, sperms or embryos), they can be passed to the offspring. The great accomplishment of CRISPR lies in the fact that it can be easily and precisely used in both reproductive and non-reproductive cells (Doudna, 2015).

Until now, altering genes in humans was limited to gene therapy in which DNA modifications were inheritable. With accuracy and efficiency of an editor that could transfer the changes to the next generation, CRISPR has renewed the debate about human germline modifications. With the advent of this technology, there was a sudden surge of genome editing experiments across the kingdom of life, but mainly restricted to organisms other than humans. Recently, there were also speculations that several scientists are secretly using this technology on human embryos. In April 2015, research involving CRISPR-based gene editing in human embryos appeared in an online journal 'Protein and Cell'. This study, done by a group of Chinese scientists (Liang *et al.*, 2015) fanned the flames of ethical controversy on human germline alterations and triggered a serious debate on issues such as how soon the technology will be able to produce designer babies. Although the experiment was done on non-viable embryos meaning that they could not develop into a live baby, it provided a proof of concept that, with little tinkering of the CRISPR enzymes, the technology might be someday successfully used on live embryos (Krishan *et al.*, 2015; Sharma and Scott, 2015). In general, apart from apprehensions related to human genome, some environmental concerns have also been identified as per CRISPR-mediated modifications in plants such as crop improvement.

Major concerns

Off-target mutations

Off-target mutations is one of the major concerns raised about the CRISPR/Cas9 genome engineering, since they may have deleterious effects on humans and the environment. Research has indicated relatively higher number of off-target mutations in human beings than in zebrafish. Larger genomes can have many identical sites for cleavage and hence CRISPR/Cas9 can cut sequences that are not intended. Such mutations can lead to an abnormal transformation and cell death. A part from that, CRISPR/Cas9 based interventions are difficult to make in cells that are hard to infect. Precise modification through CRISPR/Cas9 will play a major role in acceptance of the technology. The risks and benefits, both should be evaluated (Cong *et al.*, 2013; Hwang *et al.*, 2013; Ma *et al.*, 2014; Rodriguez, 2016; Yang *et al.*, 2013). Furthermore, such off-target mutations can become a threat to the environmental integrity. There is a possibility of transfer of genes among organisms by the ongoing process known as gene drive. In such a pattern, the off-target mutations can also be transferred to the other organisms, hence disrupting the integrity of the environment. Some scientists have warned about the risks of accidental release in the environment of experimental organisms modified using gene drive. The predecessor gene editing technologies like TALENS and ZFN need proteins designed or tailored to bind the DNA which is to be edited. These made-to-order proteins may take years to make. In comparison, CRISPR simply requires a complementary RNA to bind the target DNA taking only a few days to make. This is why CRISPR became popular as an efficient editing technique among the scientists all over the world. It is feared that with a gene editing technology like CRISPR which can make changes in DNA effortlessly, the chances of things going wrong will also be larger. For example, the plants given characteristics like drought resistance or fast growth may become invasive weeds. The most dreaded situation is that the technology may be used to produce a bioweapon such as engineering an infectious pathogen that infect humans or crops (Sarchet and Le Page, 2015). It is very important to underline the regulatory norms for the application of CRISPR. Safety and security measures should be developed to control any organism that can induce environmental damage (Oye *et al.*, 2014).

Regulatory issues

The simplicity of CRISPR/Cas9 to induce genetic modifications makes it equally difficult to identify the respective modified organism outside the laboratory and therefore raises concerns over their regulations. If CRISPR/Cas9 continues to flourish across the world, we can expect an expanded market of genetically modified organisms which will raise questions on their regulations. In addition, the patenting issues need to be resolved. There exists a friction between the scientific communities about the patenting of genetically modified organisms for therapeutic uses. Many economic interests revolve around the CRISPR/Cas9. It is assumed that patenting of CRISPR/Cas9 techniques can give enormous powers to the relevant companies (Ledford, 2015b; Rodriguez, 2016).

Genetic enhancement

For safety reasons, genetic enhancement is banned in the germlines (except in the UK where the CRISPR/Cas9 modification of human embryos has been allowed since February, 2016), however, CRISPR/Cas9 provides an easy way to manipulate the DNA sequences in the somatic cells and hence can introduce a desired phenotypic trait. For instance, one can easily improve the genetics of a sportsman. It can be used to improve the health but it may happen in the future that the criminal justice system mandates genome editing of genes related to violence for repeat offenders or violent dangerous criminals (Rodriguez, 2016). Genetic diseases that are caused by mutations in the DNA can theoretically be controlled by editing the germlines, and such modifications will be passed on to generations. Questions are raised because of the introduction of a novel change to the human genome pool (Kohn *et al.*, 2016). It will also raise an ethical debate owing a three W's pattern, like 'Who' is going to decide 'What' kind of modification for 'Whom'? Parental and guardian and extent of their authority, informed consents area some of the other ethical issues. Whether parents will be the only autonomous entity to decide for their child, or would that be tantamount to usurping the interest of future generations who cannot provide their consent at the time of the decision (Lander, 2015).

A part from that, we assume to have a significant increase in the genetically altered organisms through CRISPR/Cas9, which raises questions on the uninform level of regulations. What will happen if such modifications are being used for the non-health purpose. Societal problems will emerge if an individual or a group becomes genetically superior to other. Technology abuse and issues on the Dual Use Research Concerns (DURC) is also raised.

Patient safety

Patient safety is of vital importance among the arguments made for the acceptable application of this technology (Lanphier *et al.*, 2015). Considering the use of germline editing research in a clinical application where inheritance of a certain genetic disease may be prevented, it may relieve the parents' suffering and worries that stem from the risk of that genetic disorder in their child (Ishii, 2015). But even before a much anticipated human therapeutic application turns into a reality, it is important to proceed with extreme caution in order to avoid the undesirable effects. Recently, a synthetic biologist at Massachusetts Institute of Technology and Harvard, Feng Zhang, successfully altered the Cas9 enzyme to reduce the number of off-target mutations (Ledford, 2015a). If combined with other modifications, it is likely that the error rate may be further reduced to a safe range (Ledford, 2015a). With this possibility, the safety concerns of the CRISPR mediated germline editing may be somehow alleviated. Regrettably, even with fast pace of scientific research, the clinical application of this technology seems to be a distant dream.

Another safety concern is related to the clinical trials of CRISPR/Cas-based gene therapies. In case of chemically synthesized or natural substances and even somatic gene therapy using gene editing techniques, health safety of the subjects can be guaranteed by carefully controlling the administered doses. Also, these therapeutics degrade after a certain period of time. However, germline modifications are irreversible and may either require entirely new guidelines for the trials or improvement of existing principles adapted for somatic gene therapies in order to protect patients from adverse side effects (Baumann, 2016).

Overall, the risk potential of CRISPR/Cas technology varies for various applications. Some may be acceptable or anticipated to be applicable in the near future after addressing a few ethical and safety concerns. Others may not even be practical any time soon (Table 9.1 summarizes various potential applications and their estimated risk level). On the bright side, it may provide huge benefits in terms of improving health and environment, but it largely depends on the wise use of this technology (Sarchet and Le Page, 2015).

Ethical standpoint

The CRISPR-based gene editing experiment for human germline modification generated concerns on issues such as challenges to human safety and dignity and the risk of exploitation for eugenics. As a result, a group of stakeholders called for a voluntary moratorium on human genome research until a national or international consensus regarding the acceptance of this technology in society is reached (Ishii, 2015). The moratorium was called in a meeting held in Napa, California, with the primary goal of initiating a public debate on ethical and social impacts of the technology (Sheridan, 2015).

A variety of viewpoints about the technology in terms of its cost–benefit analysis came from a number of scientists that have been highlighted in this section. An important 'Not So Soon' opinion about the speculated use of this technology for human enhancement came from Janssens and Cecile (Janssens, 2016). They questioned the

Table 9.1 Possible risk level of CRISPR/Cas-based specific applications

Organism	Specific application	Risk level
Human	Cure diseases by replacing endogenous disease-causing genes, correcting disease causing mutations or inserting new genes with protective functions (somatic gene therapies) (Rodriguez, 2016)	Moderate (off-target mutations)
	Germline modification for introducing desired traits such as intelligence etc. or treating complex genetic disorders such as diabetes	Very high (unpredictable effects on future generations, informed consent, need to re-regularize the clinical trial procedures
Animals	Animal models for research to study diseases or development process by mutating or silencing genes. For example, a mouse model to study effect of mutations on cancer (Rodriguez, 2016)	Fairly low or no risk (less off-target effects such as in zebrafish)
Plants	Improving crops, introducing disease resistance, controlling harmful invasive species, reverse pesticide and herbicide resistance in insects and weeds through gene drive (Rodriguez, 2016)	Fairly high (off-target mutations, transfer of traits to unrelated organisms in environment, ecological imbalance due to loss of a population targeted through gene drive)

technical feasibility of CRISPR for enhancement of certain desirable traits. Practically it might be possible in case of a single variant or a limited number of variants, however, what happens when we are dealing with tens or hundreds of variants. Moreover, it is not just a matter of fine-tuning a particular gene, because complex traits such as intelligence happen due to a combination of several genes as well as their interaction with the environment, thus making the editing dangerously unpredictable in the future. This is why initial experiment on human embryos attempted to correct gene mutations in disease such as β-thalassaemia, which is a recessive genetic disorder caused by a single mutation, and not the complex diseases such as diabetes or cancer. Gene editing technology holds great promise for these type of other diseases and they estimated that tinkering with more complex conditions intended for improvement of polygenic traits or diseases will not be feasible.

According to Li and Qian (2015), CRISPR technology is not yet mature enough to fine tune human inheritance. Firstly, because of the chances of off-target mutations in the genome; secondly, it is difficult to envisage the consequences of gene editing in the next generation because all the functions of the gene may not be fully understood. However, it is said that the issues of lower efficiency may be linked to the use of non-viable embryos. And also with a little optimization of the procedure, these technical issues may be addressed (Ishii, 2015).

In all fairness, several important stakeholders recommend not to prohibit the *in vitro* germline research just on the alleged reason that the technology may be used for unethical human experimentation. Sharma and Scott (2015) support the appropriate and justified use of germline editing technology in human embryos. According to their view, as per ethical guidelines, this type of research can only be carried out on embryos before day 14 of culture. Thus, if used in accordance with the ethical procedures, there are no serious ethical concerns with using genome editing technology.

A viewpoint from Lander (2015) on the publication of genome editing techniques applied to human abnormal zygotes, strongly favoured a ban on the technology with the only exception of severe genetic diseases that have no other alternatives. He further commented that until we become scientifically more knowledgeable about the consequences of this technique, it should not be pursued on the human germline. He also pointed out the moral concerns associated with human genome editing, questioning how we would see the children as manufactured products or what would be the consequences of creating a genetic class difference by 'best genomes for the most privileged'.

Global impact of CRISPR/Cas9

CRISPR is no more a word of the scientific community. It has crossed the barriers of the laboratory into the global media and industry with largest impact on the pharmaceutical and biotech business segments. A lot of investment is pouring into multiple applications with translational medicine and specifically the gene therapy catching the most attention. A number of collaborative agreements for commercial exploitation of CRISPR have been reached between various industries, e.g. Novartis (Cambridge, MA, USA) collaborated with Intellia Therapeutics, and also signed a pact with Caribou Biosciences based in Berkeley, CA, USA (founded by Doudna) (Ricks, 2015; Sontheimer and Barrangou, 2015). With such hype and frenzy, it might not be easy to confine this technology within the boundaries of non-medical use. Thus, stringent regulations and guidelines as well as worldwide debate and awareness will pave the way for this scientific breakthrough of the new genetics era.

Way to go forward ... caution!

A solid precedent for allowing research on human germline editing has been set by the recent approval given to a team of scientists in London, UK, by the country's Human Fertilisation and Embryology Authority (1 February, 2016). The team, led by Kathy Niakan (Francis Crick Institute), plans to alter genes in healthy human embryos using CRISPR/Cas9 technology. In order to proceed the research in an ethically sound way, they will perform the experiment after fertilization and stop at the seventh day after which embryos will be destroyed (Callaway, 2016). Now that an initiative for such type of research has been taken, it calls for a world open dialogue, drafting of reports, establishment of sound policies and guidelines involving the scientific and regulatory authorities as well as the society. An important effort in this regard was a three day International Summit on human genome editing. The national academies of the USA, UK and China organized this international meeting in December 2015, which was hosted in Washington DC. The summit declared that the technology should not be used on the human germline for the purpose of establishing a pregnancy. The statement issued at the end of the summit also cautioned that ethical issues must be addressed before attempting any germline modification for clinical use. It is expected that representatives from the three countries will come up with a consensus report based on the issues discussed in the meeting by the end of 2016 (Reardon, 2015).

The risk of a powerful technology such as that of CRISPR/Cas9 lies in its users rather than the technology itself. In the past, scientists as well as the regulatory bodies have often successfully assessed both the real and seemingly apparent concerns of various scientific developments and established regulatory guidelines to keep those technologies in check. It is widely anticipated that CRISPR will be implemented only after all its risks and benefits have been carefully considered (Sontheimer and Barrangou, 2015). It is suggested that in countries where extensive legislation for human germline research exists, it might be appropriate to allow the Cas9-mediated gene editing research to proceed with proper justification. Its use may be strictly prohibited for any unethical non-medical use. However, they must discuss the social and ethical implications of this technology with the public to gain their confidence. As for the future applications of this research such as to correct genetic diseases in unborn babies, it must be strictly regulated by newly established legislation (Ishii, 2015).

In order to ensure the responsible use of this technology, germline modifications for therapeutic applications in humans should be strongly discouraged until the discussion on environmental, social and ethical concerns is going on among the stakeholders. With regard to its applications in gene therapy, a transparent research must be encouraged to assess its safety and efficacy (Baltimore *et al.*, 2015). In the long run, safety and ethical concerns associated with CRISPR technology should not halt the scientific development aimed at curing human ailments (Lanphier *et al.*, 2015).

Conclusion

Despite the fact that CRISPR genome engineering faces numerous regulatory and societal hurdles, the potential applications cannot be ignored (Fig. 9.1). CRISPR can significantly advance our understanding of diseases at genetic level while it also provides new horizons for treatment and other research applications. There is a great need to uptake CRISPR technology on various platforms that includes participation from experts from ethical, social, religious, legislative and technological

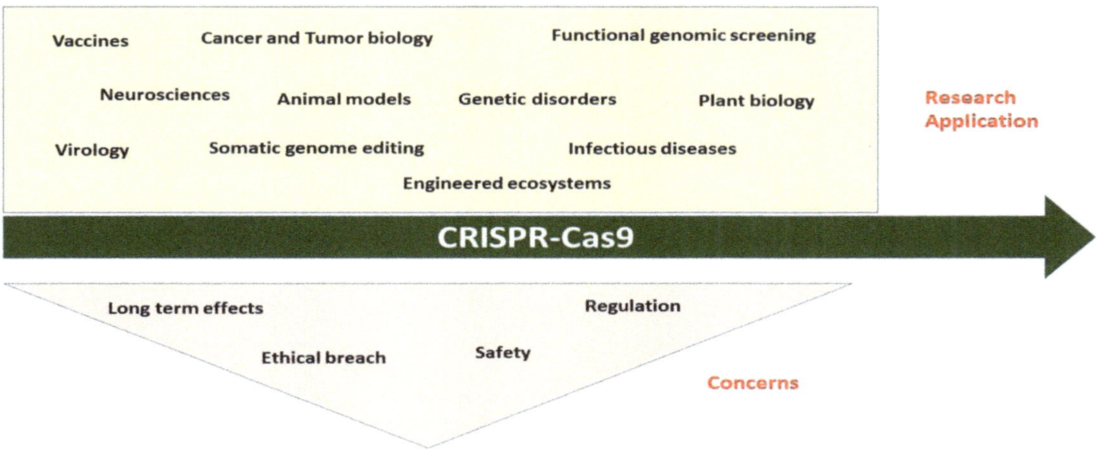

Figure 9.1 Potential applications and concerns about CRISPR.

grounds to develop a long lasting policy regarding benefits and concerns about CRISPR technology. CRISPR/Cas9 initiates many ethical and social issues not only from human perspectives but also for the environment. Risk assessments for ecological and environmental concerns should also be performed.

References

Baltimore, D., Berg, P., Botchan, M., Carroll, D., Charo, R.A., Church, G., Corn, J.E., Daley, G.Q., Doudna, J.A., Fenner, M., *et al.* (2015). Biotechnology. A prudent path forward for genomic engineering and germline gene modification. Science *348*, 36–38. http://dx.doi.org/10.1126/science.aab1028

Baumann, M. (2016). CRISPR/Cas9 genome editing–new and old ethical issues arising from a revolutionary technology. NanoEthics, 1–21.

Bolotin, A., Quinquis, B., Sorokin, A., and Ehrlich, S.D. (2005). Clustered regularly interspaced short palindrome repeats (CRISPRs) have spacers of extrachromosomal origin. Microbiology *151*, 2551–2561.

Brouns, S.J., Jore, M.M., Lundgren, M., Westra, E.R., Slijkhuis, R.J., Snijders, A.P., Dickman, M.J., Makarova, K.S., Koonin, E.V., and van der Oost, J. (2008). Small CRISPR RNAs guide antiviral defense in prokaryotes. Science *321*, 960–964. http://dx.doi.org/10.1126/science.1159689

Caplan, A.L., Parent, B., Shen, M., and Plunkett, C. (2015). No time to waste – the ethical challenges created by CRISPR: CRISPR/Cas, being an efficient, simple, and cheap technology to edit the genome of any organism, raises many ethical and regulatory issues beyond the use to manipulate human germ line cells. EMBO Rep. *16*, 1421–1426. http://dx.doi.org/10.15252/embr.201541337

Carroll, D., and Charo, R.A. (2015). The societal opportunities and challenges of genome editing. Genome Biol. *16*, 242. http://dx.doi.org/10.1186/s13059-015-0812-0

Cong, L., Ran, F.A., Cox, D., Lin, S., Barretto, R., Habib, N., Hsu, P.D., Wu, X., Jiang, W., Marraffini, L.A., *et al.* (2013). Multiplex genome engineering using CRISPR/Cas systems. Science *339*, 819–823. http://dx.doi.org/10.1126/science.1231143

Doudna, J. (2015). Perspective: Embryo editing needs scrutiny. Nature *528*, S6. http://dx.doi.org/10.1038/528S6a

Doudna, J.A., and Charpentier, E. (2014). Genome editing. The new frontier of genome engineering with CRISPR-Cas9. Science *346*, 1258096. http://dx.doi.org/10.1126/science.1258096

Guy, C.P., Majerník, A.I., Chong, J.P., and Bolt, E.L. (2004). A novel nuclease-ATPase (Nar71) from archaea is part of a proposed thermophilic DNA repair system. Nucleic Acids Res. *32*, 6176–6186.

Hsu, P.D., Lander, E.S., and Zhang, F. (2014). Development and applications of CRISPR-Cas9 for genome engineering. Cell *157*, 1262–1278. http://dx.doi.org/10.1016/j.cell.2014.05.010

Hwang, W.Y., Fu, Y., Reyon, D., Maeder, M.L., Tsai, S.Q., Sander, J.D., Peterson, R.T., Yeh, J.R., and Joung, J.K. (2013). Efficient genome editing in zebrafish using a CRISPR-Cas system. Nat. Biotechnol. *31*, 227–229. http://dx.doi.org/10.1038/nbt.2501

Ishii, T. (2015). Germline genome-editing research and its socioethical implications. Trends Mol. Med. *21*, 473–481. http://dx.doi.org/10.1016/j.molmed.2015.05.006

Ishino, Y., Shinagawa, H., Makino, K., Amemura, M., and Nakata, A. (1987). Nucleotide sequence of the iap gene, responsible for alkaline phosphatase isozyme conversion in Escherichia coli, and identification of the gene product. J. Bacteriol. *169*, 5429–5433.

Janssens, A.C. (2016). Designing babies through gene editing: science or science fiction? Genet. Med. *18*, 1186–1187. http://dx.doi.org/10.1038/gim.2016.28

Jinek, M., Chylinski, K., Fonfara, I., Hauer, M., Doudna, J.A., and Charpentier, E. (2012). A programmable dual-RNA-guided DNA endonuclease in adaptive bacterial immunity. Science *337*, 816–821. http://dx.doi.org/10.1126/science.1225829

Kohn, D.B., Porteus, M.H., and Scharenberg, A.M. (2016). Ethical and regulatory aspects of genome editing. Blood *127*, 2553–2560. http://dx.doi.org/10.1182/blood-2016-01-678136

Krishan, K., Kanchan, T., and Singh, B. (2015). Human genome editing and ethical considerations. Sci. Eng. Ethics, 1–3.

Lander, E.S. (2015). Brave New Genome. N. Engl. J. Med. *373*, 5–8. http://dx.doi.org/10.1056/NEJMp1506446

Lanphier, E., Urnov, F., Haecker, S.E., Werner, M., and Smolenski, J. (2015). Don't edit the human germ line. Nature *519*, 410–411. http://dx.doi.org/10.1038/519410a

le Page, M. (2015). Will CRISPR gene-editing technology lead to the designer babies? New Scientist. Available at: www.newscientist.com/article/mg22830500-500-will-crispr-gene-editing-technology-lead-to-designer-babies/

Ledford, H. (2015a). Biologists create more precise molecular scissors for genome editing. Nature. Available at: www.nature.com/news/biologists-create-more-precise-molecular-scissors-for-genome-editing-1.18932

Ledford, H. (2015b). CRISPR, the disruptor. Nature *522*, 20–24. http://dx.doi.org/10.1038/522020a

Li, C.X., and Qian, H.L. (2015). A double-edged sword: CRISPR-Cas9 is emerging as a revolutionary technique for genome editing. Mil. Med. Res. 2, 25. http://dx.doi.org/10.1186/s40779-015-0054-1

Liang P., Xu Y., Zhang X., Ding C., Huang R., Zhang Z., Lv J., Xie X., Chen Y., Li Y., *et al.* (2015). CRISPR/Cas9-mediated gene editing in human tripronuclear zygotes. Protein Cell *6*, 363–372.

Ma, Y., Zhang, L., and Huang, X. (2014). Genome modification by CRISPR/Cas9. FEBS. J. *281*, 5186–5193. http://dx.doi.org/10.1111/febs.13110

Makarova, K.S., Aravind, L., Grishin, N.V., Rogozin, I.B., and Koonin, E.V. (2002). A DNA repair system specific for thermophilic Archaea and bacteria predicted by genomic context analysis. Nucleic acids research *30*, 482–496.

Marraffini, L.A., and Sontheimer, E.J. (2008). CRISPR interference limits horizontal gene transfer in staphylococci by targeting DNA. Science *322*, 1843–1845. http://dx.doi.org/10.1126/science.1165771

Mojica, F.J., Díez-Villaseñor, C., Soria, E., and Juez, G. (2000). Biological significance of a family of regularly spaced repeats in the genomes of Archaea, Bacteria and mitochondria. Molecular microbiology *36*, 244–246.

Oye, K.A., Esvelt, K., Appleton, E., Catteruccia, F., Church, G., Kuiken, T., Lightfoot, S.B., McNamara, J., Smidler, A., and Collins, J.P. (2014). Biotechnology. Regulating gene drives. Science *345*, 626–628. http://dx.doi.org/10.1126/science.1254287

Reardon, S. (2015). Gene-editing summit supports some research in human embryos. Nature. Available at: www.nature.com/news/gene-editing-summit-supports-some-research-in-human-embryos-1.18947

Ricks, D. (2015). Scientists Journey Into Genomes Via CRISPR-Cas9. J. Natl. Cancer. Inst. *107*, djv352. http://dx.doi.org/10.1093/jnci/djv352

Rodriguez, E. (2016). Ethical issues in genome editing using Crispr/Cas9 System. J. Clin. Res. Bioethics *2016*.

Sarchet, P., and Le Page, M. (2015). Starting gun fired on gene editing. New Scientist *226*, 8–9.

Sharma, A., and Scott, C.T. (2015). The ethics of publishing human germline research. Nat. Biotechnol. 33, 590–592. http://dx.doi.org/10.1038/nbt.3252

Sheridan, C. (2015). CRISPR germline editing reverberates through biotech industry. Nat. Biotechnol. 33, 431–432. http://dx.doi.org/10.1038/nbt0515-431

Sontheimer, E.J., and Barrangou, R. (2015). The bacterial origins of the CRISPR genome-editing revolution. Hum. Gene Ther. *26*, 413–424. http://dx.doi.org/10.1089/hum.2015.091

Yang, H., Wang, H., Shivalila, C.S., Cheng, A.W., Shi, L., and Jaenisch, R. (2013). One-step generation of mice carrying reporter and conditional alleles by CRISPR/Cas-mediated genome engineering. Cell *154*, 1370–1379. http://dx.doi.org/10.1016/j.cell.2013.08.022

Index

www.ingramcontent.com/pod-product-compliance
Ingram Content Group UK Ltd.
Pitfield, Milton Keynes, MK11 3LW, UK
UKHW061959041026
15591UKWH00016B/1078

9 781910 190630